Internet-of-Things (IoT) Systems

Dimitrios Serpanos • Marilyn Wolf

Internet-of-Things (IoT) Systems

Architectures, Algorithms, Methodologies

 Springer

Dimitrios Serpanos
Electrical & Computer Engineering
University of Patras
Patras, Greece

Marilyn Wolf
School of ECE
Georgia Institute of Technology
Atlanta, GA, USA

ISBN 978-3-319-88828-6   ISBN 978-3-319-69715-4   (eBook)
https://doi.org/10.1007/978-3-319-69715-4

This Springer imprint is published by Springer Nature
The registered company is Springer International Publishing AG
The registered company address is: Gewerbestrasse 11, 6330 Cham, Switzerland

Dimitrios Serpanos:
To Loukia and Georgia for their patience
and understanding...

Marilyn Wolf:
To Alec.

Preface

The Internet of Things is the evolutionary step of the Internet that creates a worldwide infrastructure interconnecting machines and humans. As the Internet became public in the early 1990s, the first wave of its exploitation and deployment was mainly focused on the impact to everyday services and applications that changed the known models for financial transactions, shopping, news feeding and information sharing. It was a revolution that digitized a wide range of services as we knew them, from banking and retail shopping to face-to-face communication and government services. The first two decades of the Internet revolution focused strongly on consumer services and businesses, but human-centric. New business models appeared for banking, for online shopping, video communication, etc. for consumers. Business to business models and the cloud have impacted businesses significantly, wiping out large sectors of industry that did not adjust to the fast pace of the revolution. The impact on the economies has been tremendous. Now, more than two decades later, we witness and experience a new way of life because of the Internet's reach to our homes and work environments.

The advances of communication technology that enables the deployment and success of the Internet at home and work had an additional effect: the development of sophisticated interconnections among machines in the operational environment; we contrast the operational technology (OT) environment, which controls physical machines, to the information technology (IT) environment where humans are using computers for work. The already automated industrial environment received well the emerging technologies, adopted the suitable ones and created a, private mostly, network infrastructure that enables highly productive industrial processes. It has only been a natural step to evolve the Internet itself to include these processes. Additionally, the control models of the industrial environment, taking advantage of the smart devices –i.e. devices that include processing, memory and networking resources- that are deployed in various environments, have been extended and used in a wide variety of application domains. Conventional application domains like transportation, aeronautics, energy production and distribution, manufacturing and health adopt similar control models, exploiting smart sensors, actuators and devices that enable control automation for sophisticated applications. Critical infrastructure of countries is run using these technologies today. This emerging Internet-of-Things (IoT) is the natural evolutionary step of the Internet revolution that started about three decades ago. Importantly, IoT is building a worldwide infrastructure that will influence all facets of our life, from agriculture to mining, from health services to manufacturing and transportation. Clearly, it will provide the infrastructure over which the new emerging AI revolution will be based.

This book addresses the fundamental IoT technologies, architectures, application domains and directions. Development of a complete IoT system and service includes several components. The hardware base includes embedded processors, memories of different types, sensors, actuators, cloud servers, intermediate processing systems, network systems and gateways. The software base includes operating systems, data bases and control applications for several application domains, to the very least. The combination of hardware and software components for control applications constitutes the base for the evolution of cyber-physical systems. VLSI capabilities play a huge role in the design of IoT systems. Event-driven, distributed operation shapes the design of architectures and applications. Specialized network protocols enable efficient communication in this environment, including appropriate machine-to-machine (M2M) communication models. These technologies are emerging with constraints and restrictions for the IoT environment that are different from the typical IT environment, because of the requirements for safety, real-time responses, low power operation, etc. Security, privacy, and safety require particular attention and special techniques.

IoT is a fast-changing field. This book provides a snapshot of its current state. We continue to work in this area and hope to create updates to this book as the field progresses.

Atlanta, GA, USA   Marilyn Wolf
Patras, Greece   Dimitrios Serpanos

Acknowledgements

We would like to acknowledge M.T. Khan, K. Katsigiannis and C. Koulamas for their support and their review and comments on the drafts.

Contents

Chapter 1
The IoT Landscape

1.1 What Is IoT?

The Internet of Things (IoT) has become a common news item and marketing trend. Beyond the hype, IoT has emerged as an important technology with applications in many fields. IoT has roots in several earlier technologies: pervasive information systems, sensor networks, and embedded computing. The term *IoT system* more accurately describes the use of this technology than does *Internet of Things*. Most IoT devices are connected together to form purpose-specific systems; they are less frequently used as general-access devices on a worldwide network.

IoT moves beyond pervasive computing and information systems, which concentrated on data. Smart refrigerators are one example of pervasive computing devices. Several products included built-in PCs and allowed users to enter information about the contents of their refrigerator for menu planning. Conceptual devices would automatically scan the refrigerator contents to take care of data entry. The use cases envisioned for these refrigerators are not so far removed from menu planning applications for stand-alone personal computers.

Sensor network research spanned a range of configurations. Many of these were designed for data collection at very low data rates. The collected data would then be sent to servers for processing. Traditional sensor network research did not emphasize in-network processing.

Embedded computing concentrated on either stand-alone devices or tightly coupled networks such as those used in vehicles. Consumer electronics and cyber-physical systems were two major application domains for embedded computing; both emphasized engineered systems with well-defined goals.

Given the wide range of advocates for IoT technology, no single, clear definition of the term has emerged. We can identify several possibilities:

- Internet-enabled physical devices, although many devices don't use the Internet Protocol

© Springer International Publishing AG 2018
D. Serpanos, M. Wolf, *Internet-of-Things (IoT) Systems*,
https://doi.org/10.1007/978-3-319-69715-4_1

- Soft real-time sensor networks
- Dynamic and evolving networks of embedded computing devices

This book is primarily interested in *IoT systems*. We use this term to capture two characteristics. First, the system is designed for one or a set of applications, rather than being an agglomeration of Internet-enabled devices. Second, the IoT system takes into account the dynamics of physical systems. An IoT system may consist primarily of sensors; in some cases it may include a significant number of actuators. In both cases, the goal is to process signals and time-series data.

Interest in the Internet of Things has been spurred by the availability of micro-electromechanical (MEMS) sensors. Integrated accelerometers, gyroscopes, chemical sensors, and other forms of sensor are now widely available. The low cost and power consumption of these sensors enables new applications well beyond those of traditional laboratory or industrial measurement equipment. These sensor applications push IoT systems toward signal processing.

IoT is also enabled by the very low cost of VLSI digital and analog electronics. As we will see, IoT nodes do not rely on state-of-the-art VLSI manufacturing processes. In fact, they are inexpensive because they are able to make use of older manufacturing lines; the lower device counts available in these older technologies are more than sufficient for many IoT systems.

IoT systems must consume very little power. Power consumption is a key factor in total cost of ownership for IoT systems. Achieving the necessary power levels requires careful attention to hardware design, software design, and application algorithms.

Security and safety are key design and operational requirements for IoT systems. As we have argued elsewhere, safety and security are no longer separable problems. The merger of computational and physical systems requires us to merge the previously separate tasks of safe physical system design and secure computer system design.

1.2   Applications

IoT systems are useful in a broad range of applications:

- Industrial systems use sensors to monitor both the industrial processes themselves – the quality of the product – and the state of the equipment. An increasing number of electric motors, for example, include sensors that collect data used to predict impending motor failures.
- Smart buildings use sensors to identify the locations of people as well as the state of the building. That data can be used to control heating/ventilation/air conditioning systems and lighting systems to reduce operating costs. Smart buildings and structures also use sensors to monitor structural health.
- Smart cities use sensors to monitor pedestrian and vehicular traffic and may integrate data from smart buildings.

- Vehicles use networked sensors to monitor the state of the vehicle and provide improved dynamics, reduced fuel consumption, and lower emissions.
- Medical systems connect a wide range of patient monitoring sensors that may be located at the home, in emergency vehicles, the doctor's office, or the hospital.

Use cases help us understand the requirements on an IoT system.

Sensor network The system may act strictly as a data gathering system for a set of sensors.

Alert system Data from sensors may be gathered and analyzed. Alerts are generated when particular criteria are met.

Analysis system Data from sensors is gathered and analyzed, but in this case, the analysis is ongoing. Reports on analytic results may be generated periodically – hourly, daily, *etc.* – or may be continuously updated.

Reactive system Analysis of sensor data may cause actuators to be triggered. We reserve the term *reactive* for systems that don't implement typical control laws.

Control system Sensor data is fed to control algorithms that generate outputs for actuators.

We can identify a class of nonfunctional requirements that apply to many IoT systems. Nonfunctional requirements on the system impose nonfunctional requirements on the components.

Event latency Latency from capture of an event to its destination may not be important for batch-oriented applications but becomes important for online analysis.

Event throughput The rate at which events can be captured, transported, and processed depends on the throughput of the nodes, network bandwidth, and cloud throughput.

Event loss rate and buffer capacity In the absence of strict upper bounds on event production rates, the environment may produce more events in an interval than the system can produce. Event loss rate captures the desired capability, while buffer capacity is a more pragmatic requirement that can be directly tied to component capabilities.

Service latency and throughput Ultimately, events will be processed by services. We can also specify the latency and throughput for services.

Reliability and availability Since IoT systems are distributed, reliability is more likely to be specified over parts of the network rather than reliability of the complete system. Availability is commonly used to describe distributed systems.

Service lifetime IoT systems are often expected to have longer lifetimes than we expect for PC systems. The lifetime of the system or a subset of the system may be considerably longer than that of a component, particularly if the system uses redundant sensors and other components.

1.3   Architectures

A key aspect of IoT is *event-driven or aperiodic sampling*. Traditional digital signal processing and control assume periodic samples resulting in time-series data. However, time series consume too much power at the nodes and too much bandwidth on the network. Not all applications are amenable to aperiodic data acquisition.

Constraints on power and bandwidth also encourage distributed computing over sensor events. Relatively small processors can perform useful processing on many data streams. Recognizing interesting events using edge processing reduces the amount of network bandwidth consumed; it also reduces power consumption since wireless communication requires large amounts of power. Cloud computing-(centralized servers) or fog computing (servers closer to the edge) can be used to perform further processing on those extracted events.

1.4   Wireless Networks

Wireless networks are integral to IoT systems. Wireless network connections simplify installation and operation of wireless networks.

However, wireless networks introduce some important problems and restrictions. Radio communication requires more power than does wired communication. Some of the wireless networks used in today's IoT devices were designed for other purposes, such as telephony and multimedia. As a result, they are not optimized for event-driven communication and consume significant amounts of power in the communications protocol.

One of the ironies of IoT is that many edge devices and their wireless networks don't operate on the Internet Protocol (IP). IP introduces significant overhead with an extra level of packetization and associated processing. Many IoT devices avoid IP and rely on upstream nodes to provide them with an Internet presence.

IoT networks are typically run by noncomputer experts. IoT wireless networks must be easy to deploy and relatively self-managing.

1.5   Devices

The characteristics of event-driven systems allow IoT nodes to be relatively simple. The realities of low-power operation also push nodes toward relatively low levels of integration.

VLSI technology and Moore's law are key factors in the rise of IoT systems because they allow nodes to be manufactured extremely cheaply. Very small chips can provide enough computation, memory, and networking for useful IoT node functions. In contrast to traditional microprocessor and consumer electronic applications, where chip areas range around or even higher, chips of several square millimeters are large enough for many IoT node devices.

1.6   Security and Privacy

Security has finally been recognized as an essential requirement for all types of computer systems, including IoT systems. However, many IoT systems are much less secure than typical Windows/Mac/Linux systems. IoT security problems stem from a range of causes: inadequate security features in hardware, poorly designed software with a range of vulnerabilities, default passwords, and other security design errors.

Insecure IoT nodes create problems for the security of the entire IoT system. Because nodes typically have lifetimes of several years, the large installed base of insecure devices will create security problems for some time to come.

Insecure IoT systems also cause security problems for the rest of the Internet. IoT devices are plentiful; insecure IoT nodes are ideally suited to denial-of-service attacks. The Dyn attack [Sch16] is one example of an IoT-based attack on traditional Internet infrastructure.

Privacy is related to security but requires specific measures at the application, network, and device levels. Not only must user data be protected from outright theft, but the network needs to be designed so that less-private data cannot easily be used to infer more private data.

1.7   Event-Driven Systems

We believe that the *event* is a fundamental data type in IoT systems and that *event-driven systems* are an important structuring technique for IoT. Many of the building block technologies used for IoT today show some holdover from traditional, transaction-oriented systems. Event processing pushes us to treat time as a first-class concept and to consider the relationship between events in event sequences.

We use the term *event* more broadly then do simulation engineers. We consider events as *time-value sets*. Event-driven system simulation is widely used for modeling a wide range of engineering systems. In that context, an event is generally used to mean a change in the state of a variable. Given the decentralized nature of IoT systems, we are willing to consider stuttering – the repetition of an event value – as part of the event model. We also use events to model sampled data and time-series data. We believe that all these uses of the term event can be unified to create rich system structures.

1.8   This Book

The rest of this book describes a range of topics in IoT systems in more detail:

- Chapter 2 studies IoT system architectures, including wireless networks.
- Chapter 3 considers VLSI IoT devices. It describes the relationship between cost of ownership, power consumption, and duty cycle.
- Chapter 4 introduces analysis methods for event-driven IoT systems. These analysis methods allow us to study the memory requirements implied by event communication and processing.
- Chapter 5 describes the Industrial Internet of Things and applications of IoT systems in smart energy systems.
- Chapter 6 studies security and safety issues in IoT systems. Computer and cyberphysical system security is closely tied to safety in sensor and closed-loop control systems.
- Chapter 7 describes fuzz testing, a technique for testing the security of IoT systems. Bugs and crashes can provide exploits for attackers; fuzz testing is designed to help identify such problems.

Reference

[Sch16]   Schneier, B. (2016, October 22). DDoS attacks against Dyn. *Schneier on Security.* https://www.schneier.com/blog/archives/2016/10/ddos_attacks_ag.html

Chapter 2
IoT System Architectures

2.1 Introduction

In this chapter, we study architectures for IoT systems. We will study typical components used for networks, databases, etc.

Figure 2.1 shows the organization of an IoT system:

- The *plant* or *environment* is the physical system with which the IoT system interacts. We will use these two terms interchangeably.
- A set of *devices* form the leaves of the network. A node may include sensors and/or actuators, processors, and memory. Each node has a network interface. A node may or may not run the Internet Protocol.
- *Hubs* provide first-level connectivity between the nodes and the rest of the network. Hubs are typically run IP.
- *Fog processors* perform operations on local sets of nodes and hubs. Keeping some servers nearer the nodes reduces latency. However, fog devices may not have as much compute power as cloud servers. Fog devices also introduce system management issues.
- *Cloud servers* provide computational services for the IoT system. *Databases* store data and computational results. The cloud may provide a variety of services that mediate between nodes and users.

2.2 Protocols Concepts

Several protocols are used for data services in IoT systems.

Communication protocols may not provide sufficient abstraction for many applications. IoT systems need multi-hop, end-to-end communication. They also may exhibit complex relationships between data sources and sinks. Higher-level protocols can provide services that model more closely the needs of IoT systems. Given

© Springer International Publishing AG 2018
D. Serpanos, M. Wolf, *Internet-of-Things (IoT) Systems*,
https://doi.org/10.1007/978-3-319-69715-4_2

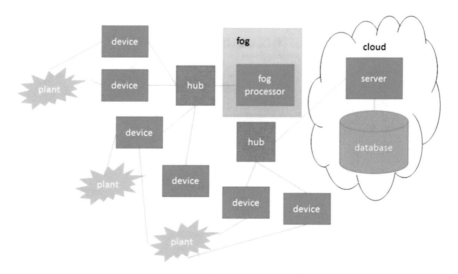

Fig. 2.1 Organization of an IoT system

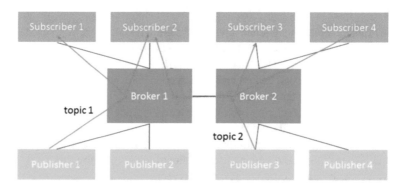

Fig. 2.2 The publish/subscribe model

the heterogeneous and long-lived nature of most IoT systems, standards are often used rather than custom protocols. Several different protocols have been proposed and, to varying degrees, used for IoT systems [Duf13]. The user space has not yet converged on a single standard for IoT communication services.

Given the prevalence of event-oriented models in IoT systems, a protocol should support event-style communication.

The HTTP protocol uses a request/response design pattern. A client issues a request for a hypertext object; the server then replies with the object in response.

A *publish/subscribe* protocol [Twi11] requires less coupling between the client and server as illustrated in Fig. 2.2. The server, known as a publisher, classifies messages into categories. Clients subscribe to the categories of interest to them. Publish/subscribe systems are typically mediated by *brokers* which receive published messages from publishers and send them to subscribers. Messages may be organized by topic; all message of a given *topic* are distributed by the brokers to the subscribers for that topic. The broker knows the identities of subscribers but the publisher does not. Brokers may interact with each other using a bridge protocol. A *bridge* allows indirect publication of messages, with a message going from the publisher to a first broker, then to a second broker, and finally to subscribers who are not connected to the first broker.

Data Distribution Service (DDS) (http://portals.omg.org/dds/) [Obj16] is a publish/subscribe software architecture; several implementations of DDS are in use. A DDS *domain* maintains a logical *global data space*; the data is managed over a set of local stores. Publishers and subscribers are dynamically discovered across the network. Publishers can specify a number of quality of service parameters that are enforced by the brokers.

Real-Time Publish/Subscribe Protocol (RTPS) [Obj14] is a so-called wire protocol that defines a protocol for communication with DDS and other publish/subscribe systems. RTPS provides QoS properties, fault tolerance, and type safety.

Esposito et al. [Esp09] developed an architecture for time-sensitive publish/subscribe systems that would be scalable to Internet-sized systems. They identified three major design goals: predictable latency, guaranteed delivery in the presence of multiple faults, and continued performance under scaling. They identified several types of fault models for publish/subscribe systems: network anomalies (loss, ordering, corruption, delay, congestion, partitioning), link crash, node crash, and churn of nodes unexpectedly joining and leaving the system. Their architecture has three abstraction layers: the network layer consists of domains composed of nodes; the nodes layer consists of clusters, with each cluster's members belonging to the same stub domain; and a coordinators layer. The coordination layer routes messages using a tree-based topology built on top of a distributed hash table. The coordinator is *p*-redundant to provide fault-tolerant coordination. To provide fault-tolerant overlays, they formulate a model for path diversity that can be computed with limited knowledge of the network connections.

Kang et al. [Kan12] used a semantics-aware communication mechanism to reduce overhead and improve reliability. They use state-space estimators at both the publisher and subscriber to maintain continuity of sensor values in the presence of network variations. Their state estimator is of the form $x_{k+1} = F_{k+1} x_k$. The designer sets a model precision bound δ for each sensor. The bound is used to manage bandwidth requirements. Their system also dynamically adjusts the model precision bound.

Choi et al. [Choi16] combined DDS with the OpenFlow software-defined networking protocol to ensure that DDS can implement the QoS parameters. They added two QoS parameters that could not be easily deduced from the standard DDS parameters: MINIMUM_SEPARATION and an E2E_LATENCY specified by subscribers.

2.3 IoT-Oriented Protocols

We can divide protocols into two major categories: those that are tied to a specific physical layer and those that are not. Generally speaking, protocols that rely on a specific physical layer do not use the Internet Protocol, while protocols that are physical layer agnostic do use IP.

Zigbee [Zig14, Far08] is a mesh network designed for low-power operation. A variety of derivative application standards specialize the protocol for applications such as smart homes and utilities. Zigbee is based on the IEEE 802.15.4 PHY and MAC standards. 802.15.4 operates in three bands: 868 MHz, 915 MHz, and 2.4 GHz. It delivers bit rates from 20 to 250 kbps, depending on the frequency band. The Zigbee NWK layer sits on top of the 802.15.4 MAC layer and provides data and management services. The APL layer includes three sections: the application support sublayer, the Zigbee Device Objects layer, and the application framework.

Zigbee provides two types of network security models: a centralized security network can be started only by a Zigbee coordinator/trust center; distributed security networks do not have a central trust center. Nodes can join either type of network and adapt to the type of network they have joined. Networks are *formed* by either coordinators or routers after scanning to select an available channel. Coordinators form centralized security networks, while routers form distributed security networks. *Network steering* is the name for the process by which a node joins a network. After identifying an open network, the node associates with that network and receives a network key. Clusters define interfaces for features and domains.

Bluetooth Low Energy (BLE) (https://www.bluetooth.com/what-is-bluetooth-technology/how-it-works/low-energy) [Hay13] is a part of the Bluetooth standard designed for low-power operation such as devices powered from coin cell batteries. A BLE device can work as a transmitter, receiver, or both. Figure 2.3 illustrates the *Bluetooth Classic* protocol stack.

The link layer provides an advertising service; devices can scan to identify nodes and networks. Devices can act as gateways to the Internet based on network address translation. The BLE protocol is stateful. BLE includes a number of optimizations to reduce power consumption.

LoRa (http://lora-alliance.org) [LoR15] is designed for wide-area IoT applications with a base station covering hundreds of square kilometers. It is designed to support a network topology with gateways for end devices, with gateways organized into their own star network. Data rates range from 0.3 to 50 kbps.

MQTT (http://www.mqtt.org) [IBM12, Oas14] is an IoT-oriented protocol with publish/subscribe semantics. The protocol is designed for low overhead and is agnostic to the data payload. MQTT provides three levels of quality of service: *at most once* provides best-effort service, *at least once* assures delivery but may incur duplicates, and *exactly once* ensures the message is delivered without duplication.

Fig. 2.3 The Bluetooth stack

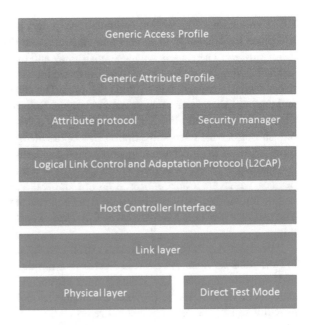

MQTT is based on a publish/subscribe model. A message is given a retention attribute when it is published; messages with QoS designations of *at least once* or *exactly once* should set the retention flag. A new subscriber to the topic will receive the last publication on that topic.

When setting up a connection, a client can provide a *will* to the server to specify a message to be published if the client is unexpectedly disconnected.

Messages are classified using topic strings similar to hierarchical file names. The set of topics is organized into a topic tree. Topic names follow the names of the nodes in the topic tree path, with node names separated by "/". Subscribers can use wildcards in the topic string: '+' denotes a wildcard match at one level of the topic tree; "#" denotes a match at any number of levels of the topic tree.

XMPP (http://xmpp.org) is a protocol for streaming XML. It provides security, authentication, and information about network availability, and rosters of clients. XMPP-IoT (http://xmpp-iot.org) is a dialect of XMPP designed for IoT applications.

REST [Vaq14, Rod15] is widely used for Web services and has received some use as an IoT service model. REST is a design pattern for stateless HTTP transfers. It exposes directory-structured form resource indicators. REST can be used to transfer XML or JSON data. Clients access resources using GET, PUT, POST, and DELETE methods.

CoAP (http://coap.technology) [IET14] is a REST-based Web transfer protocol designed for IoT devices. It can be used with several types of data payloads, including XML and JSON.

Google Cloud Pub/Sub [Goo17A, Goo17B] can be used to provide publish/subscribe service to IoT and other systems. Topics and subscriptions are exposed as

REST collections. The system is divided into a data plane for messages and a control plane for allocation to servers known as *routers*; data plane servers are known as *forwarders*. The routers balance consistency and uniformity of data using a consistent hashing algorithm. A message life cycle includes several steps. When a publisher sends a message, it is written to storage. The subscribers receive the message, and the publisher receives an acknowledgment. Subscribers acknowledge the message to Google Cloud Pub/Sub. The message is deleted from storage once at least one subscriber for each subscription has acknowledged the message. The system monitors itself to detect and mitigate service problems.

Amazon Web Services (AWS) IoT [Bar15] is a managed cloud service for IoT devices, which are termed *things*. A *thing shadow* is a cloud model of a thing. A rule engine transforms messages based on rules and routes the results to AWS services. The *message broker* is based on MQTT. A *Thing Registry* assigns unique identity to things.

Microsoft Azure (https://azure.microsoft.com/en-us/services/iot-hub/) provides IoT-oriented services. Its Service Fabric is a middleware communication system that supports microservices running on a cluster. A microservice may be either stateless or stateful. It also provides a container model for applications; a *container* provides an isolated environment but relies on the operating system, in contrast to a virtual machine which runs underneath the operating system. It provides databases using both structured and unstructured approaches. It also provides APIs for artificial intelligence services.

2.4   Databases

Databases are used for both short-term and long-term storage. Applications may rely on databases to retrieve data over a time window for analysis. Some use cases may require archival storage of values.

Unstructured databases, known as *noSQL*, are used in many IoT systems. A noSQL database does not have a schema. Simple noSQL databases represent data as key-value pairs, but other representations are possible. The lack of a schema allows quick deployment but may cause maintenance problems.

The Amazon Simple Storage Service (Amazon S3) (https://aws.amazon.com/s3/) is an object store with a Web service interface. Data can be pushed to other, lower-cost storage services for long-term, infrequent use. Notifications can be issued when objects operated upon.

Google Cloud Storage (https://cloud.google.com/storage) is an object store for unstructured data. It provides three different service models at different latency/latency/price points. Cloud SQL can be used to perform database operations. Streaming transfers are supported using HTTP chunked transfer encoding.

Time-series data possesses structure that may require special handling to provide proper database performance. Time series are sometimes stored as blobs in relational databases to allow specialized algorithms.

Dynamic time warping (DTW) [Rat04, Rak12] is widely used to search over time-series data. DTW was originally used to compare waveforms for speech processing. Correlation provides a direct comparison of two waveforms. By warping one waveform, non-exact matches can be found. Dynamic programming can be used to find the minimum warp match between two-time series; a limit on maximum warping is typically applied to avoid obviously bad matches. Very efficient algorithms have been developed to provide high-speed search. Among other techniques, these algorithms abandon a warp computation early when partial results exceed a given bound. Fast DTW algorithms have been used to search very large databases.

2.5 Time Bases

Many IoT systems require a notion of global time. Several algorithms, starting with Lamport's algorithm [Lam78], have been developed for the synchronization of clocks in a distributed system.

The Network Time Protocol (RFC1305) is used on the Internet for distributed time synchronization.

2.6 Security

Security is a system property; the system can be only as secure as its weakest component. Security features are provided by components at several layers in the IoT stack: devices, physical networks, and middleware. A unified view of IoT system security architectures has not yet emerged.

Some, but not all processors for low-power operation, provide security features such as encryption accelerators and root of trust. The National Security Agency has developed families of lightweight block ciphers [Sch13]: SIMON targets hardware implementations, and SPECK is intended for software implementations. Gulcan et al. [Gul14] developed a low-power implementation of SIMON.

Several networks provide security features. Bluetooth Low Energy provides a Simple Secure Pairing protocol to protect against passive eavesdropping. It also provides address randomization. As discussed above, Zigbee provides two network security models: centralized and distributed. LoRa provides unique network keys, unique application keys, and device-specific keys.

MQTT does not specifically require encryption, but it can be used with several different security standards. *MQTT and the NIST Framework for Improving Critical Infrastructure Cybersecurity* [Oas14B] describe the relationship between MQTT and the NIST Cybersecurity Framework.

We will study IoT system security in more detail in Chap. 6.

References

[Bar15]   Barr, J. (2015, October 8). AWS IoT: Cloud services for connected devices. *AWS Blog*. https://aws.amazon.com/blogs/aws/aws-iot-cloud-services-for-connected-devices/

[Choi16]   Choi, H.-Y., King, A. L., & Lee, I. (2016). Making DDS really real-time with OpenFlow. *2016 international conference on embedded software (EMSOFT)* (pp. 1–10). Pittsburgh, PA.

[Duf13]   Duffy, P. (2013, April 30) Beyond MQTT: A Cisco view on IoT protocols. *Cisco Blogs*. https://blogs.cisco.com/digital/beyond-mqtt-a-cisco-view-on-iot-protocols

[Esp09]   Esposito, C., Cotroneo, D., & Gokhale, A.. 2009. Reliable publish/subscribe middleware for time-sensitive internet-scale applications. *Proceedings of the third ACM international conference on distributed event-based systems* (DEBS'09). ACM, New York, Article 16, 12 pages.

[Far08]   Farahani, S. (2008). *Zigbee wireless networks and transceivers*. Amsterdam: Newnes.

[Goo17A]   Google. (2017, April 19). What is Google Cloud Pub/Sub? https://cloud.google.com/pubsub/docs/overview

[Goo17B]   Google. (2017, April 3). Google Cloud Pub/Sub: A Google-scale messaging service. https://cloud.google.com/pubsub/architecture

[Gul14]   Gulcan, E., Aysu, A., & Schaumont, P. (2015). A flexible and compact hardware architecture for the SIMON block cipher. In T. Eisenbarth & E. Öztürk (Eds.), *Lightweight cryptography for security and privacy. LightSec 2014, Lecture Notes in Computer Science* (Vol. 8898, pp. 34–50). Cham: Springer.

[Hay13]   Heydon, R. (2013). *Bluetooth low energy: The developer's handbook*. Prentice Hall: Upper Saddle River, NJ.

[IBM12]   IBM International Technical Support Organization (2012, September). *Building smarter planet solutions with MQTT and IBM WebSphere MQ telemetry*, Redbooks.

[IET14]   Internet Engineering Task Force (2014, June). *The constrained application protocol (CoAP)*, RFC 7252, Shelby, Z., Hartke, K., & Bormann, C.

[Kan12]   Kang, W., Kapitanova, K., & Son, S. H. (2012). RDDS: A real-time data distribution service for cyber-physical systems. *IEEE Transactions on Industrial Informatics, 8*(2), 393–405.

[Lam78]   Lamport, L. (1978). Time, clocks, and the ordering of events in a distributed system. *Communications of the ACM, 21*(7), 558–565.

[LoR15]   LoRa Alliance (2015, November). *LoRaWAN: What is it? A technical overview of LoRa and LoRaWAN*.

[Oas14]   Oasis. (2014, 29). *MQTT version 3.1.1*. Oasis standard.

[Oas14B]   Oasis (2014, May 28). *MQTT and the NISTG cybersecurity framework version 1.0*. Committee note 01.

[Obj14]   Object Management Group. (2014). *The real-time publish-subscribe protocol (RTPS) DDS interoperability wire protocol specification*, Version 2.2.

[Obj16]   Object Management Group. (2016). What is DDS? http://portals.omg.org/dds/what-is-dds-3/, accessed May 4, 2017.

[Sch13]   Schneier, B. SIMON and SPECK: New NSA encryption algorithms. *Schneier on Security*. https://www.schneier.com/blog/archives/2013/07/simon_and_speck.html, retrieved May 8, 2017.

[Vaq14]   Vaqqas, M. (2014, September 23) RESTful web services: A tutorial. *Dr. Dobb's*. http://www.drdobbs.com/web-development/restful-web-services-a-tutorial/240169069

[Rak12]   Rakthanmanon, T., Campana, B., Mueen, A., Batista, G., Westover, B., Zhu, Q., Zakaria, J., & Keogh, E.. 2012. Searching and mining trillions of time series subsequences under dynamic time warping. *Proceedings of the 18th ACM SIGKDD international conference on knowledge discovery and data mining (KDD'12)* (pp. 262–270). ACM, New York.

[Rat04]   Ratanamahatana, C. A., & Keogh, E. (2004, August 22–25). Everything you know about dynamic time warping is wrong. *Third workshop on mining temporal and sequential data, in conjunction with the tenth ACM SIGKDD international conference on knowledge discovery and data mining (KDD-2004).* Seattle, WA.

[Rod15]   Rodriguez, Alex. (2008, November 6). RESTful web services: The basics. IBM developerWorks, updated February 9, 2015. https://www.ibm.com/developerworks/library/ws-restful/index.html

[Twi11]   Twin Oaks Computing, Inc. (2011). What can DDS do for you?

[Zig14]   Zigbee Alliance (2014, December 2). ZigBee 3.0: The open, global standard for the Internet of Things. http://www.zigbee.org/zigbee-for-developers/zigbee/

Chapter 3
IoT Devices

3.1 The IoT Device Design Space

The design space for IoT devices is very different from that for mobile or cloud processors. Both mobile and cloud systems require very large chips. IoT devices should operate at extremely low power levels but often not operate continuously. They must integrate processors, memory and storage, communication, and sensors. They will also be sold in quantities that dwarf even those of mobile processors, which in turn require a very low price. Purchase price is, however, only one component of the IoT device cost model. Total cost of ownership will drive many IoT markets – these devices will be installed for use over a lifetime of several years. Installation cost is an important element in the decision to purchase and install these devices. We will see that cost of ownership is directly tied to power consumption.

The sensors and MEMS communities have long been interested in IoT as an application for integrated sensors and actuators. Many commentators have called for a trillion sensor world. This goal is in fact very realistic given current industry capabilities. According to Semi.org [Die16], worldwide manufacturing capacity for 200 mm wafers is expected to be 5.4 million wafers per month in 2018. If all this capacity is used for IoT, it translates to 678 billion chips per month of size 1 mm^2 or 68 billion per month of 10 mm^2 chips. That capacity puts the industry within range of producing a trillion sensors per year. We could reach the trillion sensors per year mark simply by reallocating existing capacity. Even if production does not completely reach the trillion sensor mark, the industry can clearly manufacture huge volumes of sensors.

© Springer International Publishing AG 2018
D. Serpanos, M. Wolf, *Internet-of-Things (IoT) Systems*,
https://doi.org/10.1007/978-3-319-69715-4_3

3.2   Cost of Ownership and Power Consumption

Lifetime cost of ownership is a key metric for IoT devices [Wol16]. The cost of an IoT silicon includes several components: sensing and actuation, computation, networking, as well as packaging. The completed IoT device includes power supply and packaging. However, installation cost is a significant factor in the cost of ownership. The cost of installing a cable drop in an existing building in the USA is, in the authors' experience, around $150. That cost overwhelms the cost of hardware. Eliminating all wiring – both power and networking – substantially reduces installation cost. The cost of replacing batteries is significant. Our colleague Rajesh Gupta reported that the computer science building at University of California, San Diego, requires a full-time employee to replace batteries on electronic door locks (Rajesh Gupta, personal communication, February 2014). The ability to power devices entirely by energy harvesting would eliminate that cost but imposes constraints on the devices.

The high cost and effort of wired power have encouraged the development of *energy-scavenging* (also known as *energy-harvesting*) technologies. A range of physical mechanisms can be used to convert energy for use by the environment. Since most scavenging sources provide varying amounts of power, the harvested energy is stored for later use. Electric power may be stored in a battery, a capacitor, or a supercapacitor. On-chip power management circuitry stores harvested energy and then regulates the power as it is used by the rest of the chip.

Paradiso and Starner [Par05] identified several widely different sources of energy, including radio frequency, ambient light, thermoelectricity, and heel strikes. They pointed out that indoor lighting provides much lower ambient light levels than are available from the sun. Sudevalayam and Kulkarni [Sud11] surveyed energy-harvesting technologies for sensor nodes. They identified a range of technologies with different sources, conversion efficiencies, and harvest yield. They reported, for example, that light converged by solar cells typically provided 15 mW/cm^2, wind by anemometer provided 1200mWh/day, and provided footfalls 5W.

Romani et al. [Rom17] survey power conversion and management architectures for ambient-powered IoT devices. Their reference architecture for a no-battery power management system includes several components. A transducer extracts power from an external power source with efficiency η. Several sources of internal power consumption further limit the overall system efficiency: power control circuitry consumes intrinsic power P_{int}; the storage element leaks power at a rate P_{leak}; monitor circuits consume P_{vmon}. A bootstrap circuit may be used to initialize the system from discharge. They note that a key challenge of the power management controller is to match the effective load impedance to the power source's internal impedance.

3.3   Cost per Transistor and Chip Size

Commentators have noted that established technology nodes offer cost-effective manufacturing for many products [Whi15]. One article [Hru12] quotes an NVIDIA presentation claiming that cost per transistor for 20 and 14 nm nodes was barely lower than that of the previous node and that 20 nm is "essentially worthless." Maly [Mal94] developed an early cost model for the cost of silicon as a function of manufacturing node. His model computed cost per transistor as a function of design density, minimum feature size, wafer area, and wafer cost. An even simpler cost model is based on the total cost of a manufactured wafer, including the cost of the wafer itself and all processing.

As shown in Fig. 3.1, that cost will decrease slightly as the manufacturing process matures. However, the cost of a manufactured wafer grows significantly at advanced nodes [Wol17]. Double patterning became required for lithography at 20 nm. This technique uses two masks for each feature, roughly speaking one per edge; the size of a fabricated feature can be smaller than the size of a feature on either of the masks. Double patterning requires two masks per step rather than one; since mask costs are a large part of the cost of the manufactured wafer, double patterning (and the more recent use of triple patterning) substantially increases manufactured wafer cost. Increasing the number of masks adds costs beyond those of the masks themselves: more time must be spent with the wafers in expensive equipment; wafers spend longer in the manufacturing plant.

We can write a formula for the cost per transistor based on the manufactured wafer cost C_m and the number of working transistors per wafer n_{tr}:

$$C_{tr} = \frac{C_m}{n_{tr}}.$$

In the standard Moore's Law scenario, we expect the number of transistors per wafer to double from one generation to the next. If the cost of processing the wafer increases by less than that factor, cost per transistor goes down; if not,

Fig. 3.1 Cost of processed wafers over time and technology node

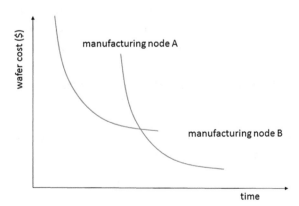

cost-per-transistor increases. Put another way, if $C_m(B)/C_m(A) > r_{tr}$ for technology nodes A and B, where r_{tr} is the factor increase in working transistors per wafer from technology A to B, then cost per transistor increases. The transition from 28 to 20 nm was an inflection point at which the cost per manufactured wafer grew enough to offset density gains. Given that these costs continued at smaller nodes, the cost-per-transistor reached a local minimum at 28 nm. It is likely that 28 nm will prove to be the global minimum of cost-per-transistor; more advanced lithography methods have their own costs.

Another major component of chip cost is silicon area. The traditional emphasis in VLSI has been on large chips to maximize functionality. However, silicon technology has advanced to the point where we can provide interesting functionality on very small amounts of silicon, thereby providing low-cost chips. This is true even for technology nodes such as 28 nm, which are large relative to the nodes used for latest-generation chips but still very dense relative to historical standards.

The transistor counts of early microprocessors provide context for the circuitry required to provide useful functionality. The IBM PC's CPU was an Intel 8088 running at 4.77 MHz [Wik16A]; the 8088 contained approximately 4000 transistors [Wik16B, Wik16C].

Packaging is another significant component of the cost of integrated circuits. A wide range of system-in-package technologies have been developed that provide several improvements over traditional single-die packaging: the ability to combine chips from several manufacturing processes, each with its own native devices; reduced inter-die parasitic values; and lower cost.

The DARPA SHIELD chip [Ral16] provides an example of a very low-cost, highly integrated IoT chip. SHIELD is designed to be attached hardware modules – chips, boards, etc. – to provide a secure, traceable identifier for that module. Since each module in a system would require its own identifier, SHIELD targets a very low manufacturing cost of one cent. The system includes several die combined in a leadless package: a CMOS module, an RF pickup coil, and a thin-film temperature sensor. The CMOS module is 100 μm × 100 μm in a 14 nm FinFET technology; it combines a digital CPU and communication, onetime programmable memory, a physically unclonable function (PUF), an analog-digital converter, and power conversion and management circuitry. The device is powered by near-field RF energy; the RF coil is used for both power delivery and communication.

3.4 Duty Cycle and Power Consumption

The *duty cycle* model is widely used to analyze IoT devices. As shown in Fig. 3.2, the model assumes periodic activation of the device. The duty cycle is the percentage of time for which the device is on:

Fig. 3.2 The IoT device duty cycle

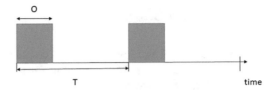

$$D = \frac{O}{T} \times 100\%.$$

Lower duty cycles mean lower energy consumption. We can change the duty cycle through a combination of changes to the operating time O and the period T. Reducing the operating time may reduce the device's functionality; increasing its period lowers its data rate.

Let the on-state power consumption of the device be P_{on}. If we assume zero leakage, then the power consumption under duty cycle operation is

$$P_{ideal} = \frac{O}{T} P_{on}.$$

If the device has a leakage power of P_{off}, then its average power consumption over the duty cycle is

$$P_{leak} = \frac{O}{T} P_{on} + \left(1 - \frac{O}{T}\right) P_{off}.$$

We can also solve for fractional duty cycle as a function of on-state and off-state power and total power consumption:

$$\frac{O}{T} = \frac{P_{leak} - P_{off}}{P_{on} - P_{off}}.$$

This model carries several implications for the design of IoT devices: the device must be good at idling at low power; it should provide low energy and time to shut down and to turn back on.

Communication power is a large fraction of the total power consumption of many IoT devices. Many IoT devices transmit small amounts of data during the on portion of their duty cycle. In this scenario, the overhead associated with setting up a communication is a significant part of the total communication power; many communication systems are designed for connection-oriented service that allows setup costs to be amortized over a longer communication.

Dementyev et al. [Dem13] measured the power consumption of several wireless protocols. They used their data to determine the optimal period T for each protocol: 14.3 s for Zigbee and 10.0 s for Bluetooth Low Energy (BLE).

3.5   Platform Design

Unlike mobile devices, most IoT devices do not operate continuously. Nonetheless, they need to retain state from activation for a range of purposes: communication status, DSP filtering, etc. SRAM requires power to retain state and thereby lengthens the allowable duty cycle. Flash memory must be written in blocks. Emerging technologies offer the promise of bit-level persistent-state devices that can be used within the processor, not just as memory.

Soerken et al. [Soe17] developed a programmable logic-in-memory (PLiM) using resistive RAM (RRAM) devices. An RRAM device has persistent state – it can be written and retains its state after the power supply is removed – making it well suited to the duty cycle characteristics of IoT devices. They designed their processor to take advantage of the majority-logic characteristics of RRAMs. They developed a compiler to translate Boolean functions into instruction streams for their processor.

3.6   Summary

IoT systems open up a new horizon for VLSI design. IoT systems require ultra-low power systems that combine disparate elements – computation, communication, and sensing – at very low price points. IoT systems emphasize small, capable chips in contrast to the large chips that have driven the industry for many years. We are at the early stages in the development of this new category of chip.

References

[Dem13] Dementyev, A., Hodges, S., Taylor, S., & Smith, J. (2013). Power consumption analysis of Bluetooth Low Energy, ZigBee and ANT sensor nodes in a cyclic sleep scenario. *Wireless Symposium (IWS), 2013 IEEE International*, Beijing, 2013, pp. 1–4.

[Die16] Dieseldorf, C. G. (2016). Foundries Take Over 200mm Capacity Fab by 2018. www.semi.org, January 25, 2016.

[Hru12] Hruska, J. (2012). Nvidia deeply unhappy with TSMC, claims 20 nm essentially worthless. extremetech.com, http://www.extremetech.com/computing/123529-nvidia-deeply-unhappy-with-tsmc-claims-22nm-essentially-worthless, March 23, 2012.

[Mal94] Maly, W. (1994). Cost of Silicon Viewed from VLSI Design Perspective. *Design Automation, 1994. 31st Conference on*, San Diego, CA, USA, 1994, pp. 135–142.

[Par05] J. A. Paradiso and T. Starner, Energy scavenging for mobile and wireless electronics IEEE Pervasive Computing, vol. 4, no. 1, pp. 18–27, 2005.

[Ral16] Ralston, P., Fry, D., Suko, S., Winters, B., King, M., & Kober, R. (2016). Defeating counterfeiters with microscopic dielets embedded in electronic components. *Computer, 49*(8), 18–26.

[Rom17] Romani, A., Tartagni, M., & Sangiorgi, E. (2017). Doing a lot with a little: Micropower conversion and management for ambient-powered electronics. *Computer, 50*(6), 41–49.

[Soe17] Soeken, M., Gaillardon, P. E., Shirinzadeh, S., Drechsler, R., & Micheli, G. D. (2017). A PLiM computer for the internet of things. *Computer, 50*(6), 35–40.

[Sud11] Sudevalayam, S., & Kulkarni, P. (2011, Third Quarter). Energy harvesting sensor nodes: Survey and implications. *IEEE Communications Surveys & Tutorials, 13*(3), 443–461.

[Whi15] White, M. (2015). IoT, Cost-per-Transistor Extend Lifetimes of Established Technology Nodes. *Electronic Design,* May 15, 2015, http://electronicdesign.com/eda/iot-cost-transistor-extend-lifetimes-established-technology-nodes

[Wik16A] Wikipedia. (2016). IBM Personal Computer. https://en.wikipedia.org/wiki/IBM_Personal_Computer. Accessed October 19, 2016.

[Wik16B] Wikipedia. (2016). Intel 8088. https://en.wikipedia.org/wiki/Intel_8088. Accessed October 19, 2016.

[Wik16C] Wikipedia. (2016). Transistor count. https://en.wikipedia.org/wiki/Transistor_count, Accessed October 19, 2016.

[Wol16] Wolf, M. (2016). Ultralow power and the new era of not-so-VLSI. *IEEE Design & Test, 33*(4), 109–113.

[Wol17] Wolf, M. (2017). *The physics of computing*. Cambridge MA: Elsevier.

Chapter 4
Event-Driven System Analysis

4.1 Introduction

This chapter describes modeling and analysis methods for Internet of Things (IoT) system design. IoT systems require new types of analysis because events do not necessarily result in immediate actions or maintain their order relative to other events.

Traditional methods such as the distributed control-oriented methods of Thiele and Ernst consider possibly infinite streams of events or samples, but the lifetime of an event/sample in the system is relatively short. In contrast, IoT systems must deal with event lifetimes at multiple time scales: some events may schedule activity only seconds in the future, while other events may schedule activity days, weeks, or months ahead. IoT also do not maintain temporal order of causality – one event may cause an event in the near future, while another event may cause an event in the far future. We need new analytical methods for multiple time scales and complex causality relationships.

The primary goal of our analysis is the understanding of the required characteristics of the IoT platform. We propose a model of the IoT system as a network with devices as leaf nodes and hubs as non-leaf nodes. Hubs perform routing functions but for our purposes their key role is to control the timing of event activity through the use of *timewheels*. While we assume that events carry key-value pairs, we are not concerned here with the semantics of events. Instead, we analyze the lifetimes of event populations. Event populations depend in part on the activity of the environment in which the IoT operates. To accommodate a wide range of realistic scenarios, we develop models based on both deterministic and stochastic event timing.

© Springer International Publishing AG 2018
D. Serpanos, M. Wolf, *Internet-of-Things (IoT) Systems*,
https://doi.org/10.1007/978-3-319-69715-4_4

4.2 Previous Work

Several lines of work have established event-based models for real-time networked systems. One of the goals of these projects has been to unify the analysis of network-oriented events and the computation on the network nodes that transform one stream of events into another.

Chakraborty et al. [Cha03] developed a real-time calculus that models events and resources. They model an event stream $R[s, t]$ over the prescribed time interval as a pair of arrival curves: $\alpha^l(\Delta)$ for the lower bound on the number of events in the interval Δ and $\alpha^u(\Delta)$ for the upper bound of events in the interval. They show how to model event streams with jitter. They use β functions to model the service provided by computational and communication components. They show how to analyze single streams, multiple interacting streams, and platforms with multiple computing and communication resources. Maxiaguine et al. [Max04] used work-load curves to characterize the computational workload of real-time systems. They showed how to use their methods to analyze both a rate-monotonic system and streaming architectures.

Henia et al. [Hen05] give definitions and formulas for events and event streams. Many of their results apply to our model; we summarize some of their applicable results here.

Event time applies to both generation and release time. An event time includes a nominal time and jitter:

$$\langle T, J \rangle$$

A periodic event stream has parameters period and jitter:

$$\langle P, J \rangle$$

The upper event function $\eta^u(\Delta t)$ gives the maximum number of events in the interval Δt. Similarly, the lower event function $\eta^l(\Delta t)$ gives the minimum number of events in the interval Δt. The upper and lower event functions for a periodic event stream with jitter are

$$\eta^u_{P+J} = \frac{\Delta t + J}{P}, \quad \eta^l_{P+J} = \max\left(0, \frac{\Delta t - J}{P}\right)$$

They give formulas for the jitter of the output of components that combine event streams using AND and OR combination methods.

4.3 Motivating Example

IoT systems are built for a variety of applications: industrial control, environmental monitoring, logistics, etc. We will use examples in this paper derived from our experiments with IoT systems for long-term care [Wol15]. This application provides us with use cases typical of smart homes (turning on and off lights, energy management, etc.) as well as use cases associated with health care (scheduling medications, checking on the condition of residents, etc.). Our example IoT system operates in a home with several residents, as a rotating set of staffers, and visitors. A variety of sensors monitor activity in the home: cameras, utility sensors, smart objects, etc. The IoT system is designed to track the activity of residents and staffers and to alert staffers of situations that may deserve their attention.

The system architecture consists of several elements:

- A set of sensors
- A local hub that monitors the sensors as I/O devices
- A cloud-based node for some analytical functions

A key feature of the local hub is its internal *timewheel* (Coelho, D., 2014, August 2, private communication). Timewheels are used in event-driven simulation to manage simulator event activity; in this case, we use the timewheel to manage events in the real world as mediated by the I/O devices. Events are timestamped with a time at which they should occur, which may be later than the time at which the event was generated. The timewheel is a time-sorted queue; when the clock time equals the timestamp of the event at the head of the timewheel, that event is dequeued and processed.

4.4 IoT Network Model

Our IoT network model is oriented toward the analysis of event behavior in the system. Because events have long lives, memory in the form of timewheel queues plays a critical role in the model.

4.4.1 Events

The model of computation is based on long-lived events. An event is generated at a node and stored until it is ready to be completed, at which time it is consumed by a node.

An event is a 5-tuple:

$$\langle key, value, dest, gen\_time, release\_time \rangle$$

The semantics of the event is given by the key-value pair. The destination of the event is the device that should process the given key-value pair. The modeling methods described in this paper are not concerned with the semantics of key-value pairs.

We also need to know the temporal behavior of an event, which is given by two values. The *generation time* ϑ is the time at which the event was created. The generation time is useful in our analysis; an implementation may or may not keep track of this value. The *release time* ρ of an event allows the IoT system to perform delayed actions – one event in the environment may not cause an immediate response but rather one that happens some time later. We refer to the difference between generation and release time as *lifetime* of an event is $\lambda = \rho - \vartheta$. Events may be generated periodically or aperiodically. Activation or release times may be periodic or aperiodic.

4.4.2   Networks

A network consists of *nodes* and *links*. We will discuss nodes in more detail below.

We model communication links are unidirectional. Most physical hubs are full duplex, but we model links as unidirectional to advance our analysis.

4.4.3   Devices and Hubs

A node may be one of two types: *device* or *hub*.

A device appears only as leaves in the network. A device has at most one input and at most one output link. Logically, a device receives or generates events. Physically, event generation can be caused either by physical events or by internal node activity; physical event receipt may cause the physical node to initiate a physical event or change its internal state.

Hubs are non-leaf nodes in the network. A hub may have more than one input or output link. Hubs may include computing and storage. However, for analytical purposes, the key role of a hub is to sequence events. A hub maintains a timewheel, a time-ordered queue of events, and a clock. (The timewheel is a notion borrowed from an event-driven simulation, but we use it in this case to manage events in cyber-physical systems.) Events arriving from the hub's devices are entered into the timewheel in order of activation time. The head event of the queue is removed from the timewheel when its activation time equals the clock time. The hub then sends the event to its destination.

Hubs must keep track of real time in order to dispatch events. Devices may or may not need to keep track of real time, depending on their functions. Given that events are dispatched by the hubs, they can rely on the hub's notion of real time to initiate events. When scheduling events, they may be able to set the release time for the event relative to the current time, which avoids the need for the device to directly keep track of real time. Our model only assumes that hubs are required to know real time, which simplifies analysis. Allowing devices to avoid maintaining real-time clocks may have some advantage in implementation as well.

4.4.4 Single-Hub Networks

A single-hub network consists of one hub mode and one or more device nodes and their associated links. The hub manages the exchange of events between its device nodes.

In a single-hub network, input traffic arrives at the hub from its device nodes, while output traffic is generated by the timewheel and goes to the devices.

As a simple example, consider scheduling medications for residents of the home. If a resident receives medicines twice per day, once in the morning and again in the evening, the device responsible for scheduling the medicines must generate an event for each administration. The event for the next medicine administration is probably generated when the current medicine administration is released, giving an event lifetime of 12 h.

The morning routine of the residents presents a more complex set of events and more scheduling choices. Each resident's routine will generate a series of events (getting up, toileting, eating breakfast, etc.); depending on the activity, all the events in the routine may be scheduled at once, or some may be scheduled on the completion of other events. If all residents get up at once, they create both congestion in the house and congestion in the hubs and their timewheels – the maximum number of events in the system will be a function of the number of residents as well as the complexity of their routines. By staggering the timing of their activities, we can both reduce physical congestion as well as reduce the number of events that the hubs must deal with at any given time.

4.4.5 Multi-hub Networks

A more general network may contain more than one type of hub. One link or a pair of links is used to connect the hubs. For the moment, we consider only tree-structured networks.

In our example system, the in-house system consists of a hub and a set of devices. The cloud analytics system also uses a hub and timewheel to manage the times at which events should be processed. For modeling purposes, the analytics engine itself is a device.

We model event routing as hub-to-hub transfers in which an event is removed from one timewheel and placed on another. When an event is transmitted to another hub, we may use additional queue operators to remove the event before it reaches the head of the queue. We will discuss the effects of event routing in more detail in the next section.

4.4.6   Network Models and Physical Networks

The mapping between model nodes/links and nodes/links in the physical network need not be one-to-one. A single physical device may house several logical nodes. A single network physical link may be used to transport several logical links.

We can rely on results from parallel computing [Dua02] for techniques for routing events in multi-hub networks. Physical networks may use separate memories for queues and buffers on network links.

The network model helps us to understand the behavior of more complex physical links. We can first separately analyze half-duplex traffic on links and then use that analysis to understand the characteristics of full-duplex links.

4.5   IoT Event Analysis

The theories for event-based analysis of distributed control networks described in Sect. 4.2 were designed for transducer networks in which events maintain their time order. In contrast, our events may be generated in one order but released in another order. The reordering effects of release times and the timewheel substantially change the analysis of IoT networks as compared to distributed control. We start with analysis of event populations using simple models of event generation. We then go on to identify stochastic models that are useful for the analysis of IoT event systems.

4.5.1   Event Populations

Because events in IoT systems are long-lived, we must consider the lifetimes of event populations. Because events may be released long after they are generated, the system may need to accommodate a large number of events even if no events are currently being generated.

The event population is the number of events that are still alive, given by the difference between the number of generated and released events. We can evaluate event population over the entire network or over a set of components in the network. When events are generated and released with jitter, we can write formulas for the upper and lower population; here we concentrate on a jitter-free form of the analysis to emphasize basic principles.

A general form for the population count is to enumerate all events from the system start time:

$$P(t) = \int_0^t \left[\vartheta(t) - \rho(t) \right] dt.$$

This formulation is cumbersome since it requires the entire system history. However, without some knowledge of the event lifetimes, we can do no better. And without a bound on event lifetime, the number of events in the system can increase without limit.

A practical consequence of this observation is that useful IoT systems must put an upper bound on the lifetimes of events.

If we have a maximum lifetime L on the lifetime of an event, we can write the event population as

$$P(t) = P(t-L) + \int_L^t \left[\vartheta(t) - \rho(t) \right] dt$$

We can also write a version of this equation taking into account event jitter.

We need to know the event population at time $t-L$ because some events may have been generated that have not yet expired.

If no events are generated in an interval L then we can guarantee that the event population at the end of that interval is zero.

Event population determines buffer requirements for components. The maximum population determines the memory requirements of the timewheel queue. Maximum populations on links help to determine the queue sizes on those links.

As a simple example, consider a single-hub network. The event stream controls medication dispensing, with medications being dispensed every 12 h. Scheduling medications may be done separately from dosing them, but let us assume for the moment that each medication dispensation also causes the next dispensing event to be scheduled. If we let $T = 12$ h and assume that one person is in the system, then $\gamma^u(T) = \gamma^l(T) = 1$, $\alpha^u(T) = \alpha^l(T) = 1$, and $\eta^u(T) = \eta^l(T) = 2$. The maximum lifetime is $L = 12$ h. The event population at $t = 12 - \epsilon$, just before the first set of prescription dispensed is released and the second set generated is

$$P(12 - \epsilon) = \int_0^{12-\epsilon} \left[\vartheta(t) - \rho(t) \right] dt = 2$$

We can easily generalize this formula to the case of n people.

The maximum population in an interval $[t_1, t_2]$ is

$$P_{max}(t_1, t_2) = \max_{t \in [t_1, t_2]} P(t)$$

Event generation in many IoT systems is not strictly periodic – some events or event streams may be activated aperiodically. In this case, the event population depends on the use case.

We can evaluate event populations when event characteristics are stochastic. For example, consider an event stream that is generated periodically at a rate of one per second. The activation times (measured relative to the generation time) are given by a uniform distribution over the interval $[1, 10]$. The maximum population is

Fig. 4.1 A multi-hub
network

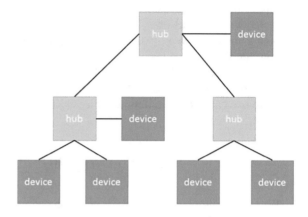

$$P_{\max}\left(t_1,t_2\right) = \max_{t\in[t_1,t_2]}\int_{t_1}^{t_2}\left[\vartheta(t)-\rho(t)\right]dt$$

We will develop in Sect. 4.5.3 techniques to characterize event populations under several different models of event generation.

For the moment, consider the case of the morning routine for n people. Let us assume for concreteness that a stream of four events is generated, each 1 min apart, with events released 5 min after generation. If everyone gets up at once, then the event population in the first 4 mins is $P(4)=4n$ and $P_{\max}=4n$. If we stagger the schedules of the residents so that each gets up 5 mins apart, then the maximum population reduces to 4.

In the case of a multi-hub system, different nodes may have different event populations and different maximum event populations. In the smart home, we perform some operations locally and some in the cloud. We model this with the network shown in Fig. 4.1: two hubs are in the home, one for input devices and one for output devices (a choice made here for modeling clarity); one hub is in the cloud. The cloud hub is connected to a single device that performs analysis algorithms. The analysis algorithms consume events, process them, and then possibly generate output events. (One example of such analysis is tracking [2].)

4.5.2 Stochastic Event Populations

A wide variety of assumptions and stochastic models are possible for events. In this section, we use some basic models to derive important design metrics. Although no one to our knowledge has gathered large traces of IoT activity, we can use models of traffic from related domains to help us understand IoT design.

We can gain some intuition by considering the simpler case of the Poisson distribution. A common model for telephone traffic is that call arrivals and departures are each modeled as Poisson processes. In our case, we use the Poisson distribution to model event generation at a rate λ. If successive events have non-overlapping lifetimes, then their maximum population in that interval is 1; if their lifetimes overlap, then the maximum population is 2, which must be accommodated by buffering. The probability that two events have overlapping lifetimes L is

$$P[t < L] = 1 - e^{-\lambda L}$$

This simple formulation suggests that λL, the product of event generation rate and lifetime, is a useful metric for judging maximum event populations.

The Erlang-B distribution provides a more accurate model for event populations. In the case of IoT events, the event dwell time corresponds to call duration; the queues correspond to telephone lines. (The Erlang-C distribution models call waiting with queues. In our case, consider the queue as a set of servers consisting of memory locations. One memory location/server is required for each event to wait for its release time.)

The offered traffic is in units of erlangs:

$$E = \lambda L$$

In our case, λ is the event arrival rate and L is the event lifetime. The probability of blocking (i.e., dropping an event due to a full queue) is

$$P_b = B(E,m) = \frac{E^m / m!}{\sum_{i=0}^{m} E^i / i!}$$

where m is the size of the queue.

The offered event traffic in erlangs is a useful rule-of-thumb metric for IoT system traffic – both the frequency of events and their dwell times must be considered to understand their effect on timewheel size.

We can use P_b to design the timewheel capacities of the hubs, either using the maximum traffic as a guide or evaluating the traffic at different points in time using the population functions. Given the systemwide offered traffic, we can find P_b for the entire network. However, in a multi-hub system, we must determine how to partition the timewheel memory between the hubs.

We can model the traffic hub by hub:

$$\sum_{i=1}^{n} E_i = \sum_{i=1}^{n} \lambda_i L_i$$

From this, we can determine the P_bs. However, this approach does not minimize total system memory. If we assume that all the hubs share the same values for arrival rate and event lifetime, then $E < \sum_{1 \le i \le n} E_i$.

We describe in Sect. 4.5.4 how to transfer events between hubs to balance queue sizes.

4.5.3 Environmental Interaction Modeling

We can identify three methods for modeling the interaction of the IoT system with its environment, each with its own degree of accuracy and detail.

The simplest model treats both the device and the user as timed finite-state machines. Given a path through the user machine that defines a given use case, we can form the product of the device machine and the user machine path. The resulting FSM, along with a timing regimen that is specified by the use case, tells us when events are generated by the device. That trace can be used to build the event population trace.

A more sophisticated model treats the user as a Markov decision process (MDP) with fixed timing. A Markov decision process is a stochastic model used for optimization. An MDP is defined by a set of states and possible actions out of each state. Each action is assigned a reward R. Transitions out of the action to the next state are assigned probabilities. We can use any of several different algorithms (dynamic programming, linear algebra) to find the path that maximizes the reward. In this scenario, we solve for the optimal reward path and form its product with the device model, using a fixed time model. Figure 4.2 shows an example of a simple device model and user model. The device model combines the actions of all the component devices related to the routine into a single state machine for simplicity. The actions in the user model MDP correspond to states in the device model.

A yet more complex model uses a continuous time Markov decision process (CTMDP). The most common mathematical form of this model is as an MDP with the timing of state transitions modeled as a Poisson process [Buc11]. Standard MDP approaches can be used to solve for the optimal path with timing given by the Poisson process.

4.5.4 Event Transport and Migration

An event does not necessarily have to be stored on the hub that owns either the event's source or destination. In a multi-hub system, we can station events at nonlocal hubs to avoid overflowing a hub's queue capacity or improve its battery life. If an event is queued nonlocally, we must factor transmission time into its release to ensure that it reaches its destination device at the proper time.

For simplicity, we consider the case of no energy cost for transporting events across the network. Let P_e be the power consumption of storing one event in memory for a unit time. Given a population of events Π, the energy required to store all events in the population until their release times is

$$E_{pop} = \sum_{e \in \Pi} \left[\rho(e) - \vartheta(e) \right] P_e$$

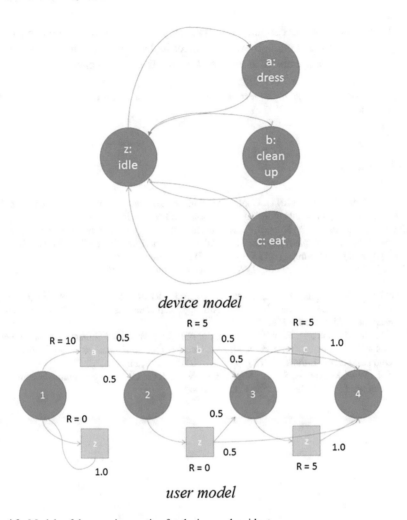

Fig. 4.2 Models of the morning routine for devices and residents

We have a set of H hubs each with available battery energy $E_h(i)$. We want to find an allocation of events to hubs such that

$$\forall i \in H : E_{pop}(i) \leq E_h(i)$$

This is a classic bin-packing problem, although we want to solve it as a distributed problem without a centralized list of events. In practice, transmission energy reduces the set of plausible event allocations.

We propose a heuristic algorithm for event migration:

- Find a partial ordering of the hubs such that no two adjacent hubs are in the same set and all hubs are covered.

- Hubs proceed in order so that no two adjacent hubs off-load simultaneously.
- Each hub off-loads enough events to meet its battery requirements. Events are moved to adjacent hubs with the greatest available battery capacity.

Acknowledgment   Thanks to the team at Alya Networks for useful discussions on key-value-based IoT networks.

References

[Buc11]   Peter Buchholz (2011). Continuous time Markov decision processes: Theory, applications, and computational algorithms. TU Dortmund Informatik IV lecture notes.

[Cha03]   Chakraborty, S., K¨unzli, S., & Thiele, L. (2003). A general framework for analysing system properties in platform-based embedded system designs. *Proceedings sixth design, automation and test in Europe (DATE)* (pp. 190–195). Munich, Germany.

[Dua02]   Duato, J., Yalamanchili, S., & Ni, L. (2002). *Interconnection networks.* San Francisco: Morgan Kaufman.

[Hen05]   Henia, R., Hamann, A., Jersak, M., Racu, R., Richter, K., & Ernst, R. (2005). System level performance analysis: The SymTA/S approach. *IEE Proceedings: Computer Design Techniques, 152*(2), 148–166.

[Max04]   Maxiaguine, A., Kunzli, S., & Thiele, L. (2004). Workload characterization model for tasks with variable execution demand. *Proceedings of the conference on seventh design, automation and test in Europe (DATE)* (pp. 1040–1045). Paris, France.

[Wol15]   Wolf, M., van der Schaar, M., Kim, H., & Xu, J. (2015). Caring analytics for adults with special needs. *IEEE Design & Test, 32*(5), 35–44.

Chapter 5
Industrial Internet of Things

5.1 Introduction

The Internet of Things (IoT) has already brought a revolution to our understanding of applications in a wide range of human activity. This trend is expected to increase in the near future, as the potential economic impact of IoT is expected to be between 900 billion USD and 2.3 trillion USD on a yearly basis up to 2025 [Man13]. IoT applications are spreading to various sectors including smart energy, manufacturing, agriculture, health, security and safety, smart cities, smart buildings, and smart environment. All these application areas repeat the same basic model: a large number of smart devices, interconnected over wired or wireless media, interacting and coordinating to achieve a goal.

In the industrial environment, the effort for smart factories [Zue10], the Industrie 4.0 strategy [Ind14], the Industrial Internet [GE17], and the European initiative for the Factories of the Future [FoF] have initiated the adoption of IoT in industry with the goals of increasing flexibility and productivity, while reducing production cost. The developing concept is the Industrial IoT (IIoT).

The Industrial Internet of Things is part of the general IoT evolution. However, it faces challenges that are unique and differentiate it from the other systems and services of IoT due to the need to integrate programmable logic controllers (PLC) and supervisory control and data acquisition systems (SCADA). PLC and SCADA systems, together with the related industrial networks that interconnect them, constitute the infrastructure of operational technology (OT), which has traditionally evolved independently from the typical IT technology, because it addresses the needs of systems in the field – industrial floor, energy production facilities, energy distribution networks, etc. – with strong requirements such as continuous operation, safety, real-time operation, etc. The capabilities offered by the emerging IIoT technology pose challenges for the integration of these OT systems with the traditional enterprise IT systems at many levels, from enterprise management to cyber security. For example, enterprise resource planning systems (ERP) need to be expanded to

© Springer International Publishing AG 2018
D. Serpanos, M. Wolf, *Internet-of-Things (IoT) Systems*,
https://doi.org/10.1007/978-3-319-69715-4_5

include manufacturing operations, which are managed currently by manufacturing execution systems (MES) that have grown independently and present significant interoperability challenges to their integration. Clearly, an integrated system that manages the complete enterprise/factory hierarchy, from business processes to sensors, provides significant flexibility and presents new opportunities to enterprises.

Industrial technology is not part only of manufacturing or factories. The maturity of the technology and its cyber-physical control capabilities has spread its use outside traditional factory environments, and now they constitute a significant part of the critical infrastructure at many fronts. Energy production and distribution infrastructure includes OT systems, which are the indispensable infrastructure on which modern smart grids are built. Actually, the energy sector is a high priority in the evolution of IIoT, not only because there is increasing need to consumers for energy, especially in light of the population growth, but also because energy management is a critical factor in the industrial sector and the desired low-cost production of goods and services. In addition to the energy sector, industrial systems are widespread in many other sectors of critical infrastructure, such as water management and transportation.

The interoperability challenges to the convergence of IT and OT are only a part of the challenges in the emerging IIoT. Appropriate architectures need to be developed to build and manage effective IIoT systems, technologies for the design and management of cyber-physical systems, sensors and networks need to be developed, and, importantly, safety and security need to be addressed in a unified way in the context of IIoT. Safety and security are significant challenges, because, traditionally, security has been a concern in the IT sector, while safety has been the major concern in the OT sector. Bringing the two together has brought the realization that safety cannot be achieved without security, while, at the same time, security needs to include technologies that combine dependability and meet strong requirements for real time and low power in many application domains. Although the security issues of industrial control systems have attracted attention in the last decade and standards have been evolving at a much faster pace than in the past, e.g., the ISO/IEC 27000 and the ISA/IEC 62443 families of standards [IEC16, ISA16], there are still significant challenges at the technology, architecture, and management fronts to obtain solutions for the unified IIoT.

In this chapter, we present the concepts and evolution of the IIoT starting from the Industrie 4.0 strategy and proceeding to the Industrial Internet. We describe the IIoT reference architectures as they evolve from the ITU effort to the Industrial Internet Consortium. Finally, we describe some representative challenges in the evolution and implementation of IIoT focusing on the energy sector. As security and safety constitute a significant challenge in IIoT as well as in IoT, in general, we focus on this challenge in the following chapter.

5.2   Industrie 4.0

Industrie 4.0 is a strategic initiative in Germany that targets to bring IoT technologies to the manufacturing and production sectors [Ind14].The goal is to enable Germany to keep a leading role in manufacturing achieving efficient and low-cost production with flexible workflows. The means to achieve this goal is the widespread inclusion of cyber-physical systems in the manufacturing and production processes, in order to insert intelligence in the systems and processes, to enable their high connectivity and communication, and to achieve their coordination into more complex but flexible processes that lead to high-quality, low-cost products.

Industrie 4.0 takes its name from the identification of the new, emerging industry as the fourth revolution of industrial production. It is widely accepted that industrial production to date has gone through three (3) revolutions. The first industrial revolution, between the eighteenth and nineteenth century, is the one where mechanized production facilities were introduced in the production of goods and services, where the required energy was provided by water and steam. Electrical energy was introduced during the second revolution, which led to mass production, as electricity boosted productivity. In the post WWII era, the inclusion of electronics and software, i.e., industrial information technology, to the mechanical and electrical components led to the third revolution that enabled automation at high levels. Currently, many industrial stakeholders believe that we are at the verge of the next, the fourth, industrial revolution, through wide adoption and use of cyber-physical systems that leads not only to even higher levels of automation but enables mass customized manufacturing and production of goods and services, due to the flexibility offered by the easily programmable, configurable, and controllable manufacturing lines.

The effort for Industrie 4.0 is based on the widespread deployment and use of computational and communication resources. The last two decades have been characterized by significant advances in high performance, low-power processors, memories, and communication components that enable efficient processing and networking. These advances have brought significant processing capabilities to a large number of devices that are deployed to consumers or to the field. Smart consumer devices have become norm. Smartphones provide hundreds of applications and enable services ranging from identifying travel and transportation routes to mobile banking and health monitoring. Smart televisions combine and provide various types of entertainment and network services, from customized TV channel control and management to Internet gaming and home device management. Smart home appliances monitor parameters, from environmental temperatures to water and energy consumption, enabling citizens to manage their homes efficiently and effectively leading to the required living quality while reducing operational cost at various fronts.

The large basis of computational resources and connectivity becomes apparent by the published numbers of embedded processors and components that are currently produced. According to [Ind14], the vast majority of produced processors, approximately 98%, are deployed in embedded systems. Deployed semiconductor memory is also growing and expected to grow at 40% year over year in 2017 [Mic17]. Furthermore, the significant advances of wired and wireless networks in the last two decades have led to ubiquitous connectivity, approaching 100% in cities and towns, through different technologies.

The available processing and communication basis leads to an evolving hierarchy of embedded systems and services up to the level of the Internet of Things, Data, and Services. Examples of this evolution can be identified at several application areas. In transportation, for example, embedded systems are widespread controlling functions from car entertainment systems to car seat control. At this level, embedded processors are programmed to control specific, individual parameters, e.g., height and movement in car seats, based on user commands. However, embedded systems in cars are also networked, either within the car system or with the environment, providing networked embedded services; automatic toll payment is one of them where embedded systems in the car and the toll booths communicate with each other, in order to complete the electronic payment transaction of the toll passage. Such payment systems from several tolls, for example, can be further combined in a distributed system that enables traffic and toll management at a wider scale, leading to more effective transportation infrastructure that achieves lower waiting times and fuel costs for travelers as well as lower operational cost and, thus, higher income to transportation management authorities. One can even envision an even higher level of connectivity of such complex transportation systems to smart cities that combine transportation management with additional services, such as energy distribution, civil services, emergency services, etc., as required at different times, locations, and during special events.

The advances of sensor technologies, in addition to the evolution of embedded systems and communication networks, make all these scenarios realistic. Importantly, sensors bridge the gap between the physical world and the digital world, providing increasingly rich information to digital systems and enabling intelligent control of systems and processes. In that respect, manufacturing and industrial automation has been traditionally employing IT technologies with sensors and electromechanical systems, leading the development and deployment of technologies and concepts for intelligent control, systems, and services. Thus, the development of the Industrie 4.0 strategy and the related initiatives comes as a natural evolution step of industrial technologies influencing and being influenced by the advancement of consumer technologies of the Internet of Things.

The smart factory concept embodies the goals of the Industrie 4.0 strategy to a large degree. The concept is based on the hierarchy of cyber-physical systems mentioned above, where smart production systems are interconnected in a multilevel hierarchy to achieve a high degree of automation, targeting flexibility, efficiency, autonomy, resilience, safety, and low cost. Smart machines will be interconnected to establish smart plants, which, in turn, will be combined to provide smart factories. Considering the typical components of manufacturing process, smart factories are targeted to automate efficiently all components and stages. Materials and resources will be managed and introduced in the process efficiently; production processing will be managed in real time minimizing the used resources for the products and the operations, while reconfiguration and reorganization of production processes and customization of products will be feasible in real time and with safety for infrastructure and operators, minimizing environmental impact. Customers will be able to monitor the progress of the development of ordered products, while manufacturers will be optimizing their logistics chains.

5.3   Industrial Internet of Things (IIoT)

The Industrial Internet of Things (IIoT) has emerged as a general concept of the application of the Internet of Things to the industrial sector. Effectively, it is a generalization of Industrie 4.0, which appears to focus more on industrial process efficiency. The IIoT vision includes all aspects of industrial operations, focusing not only on process efficiency but also on asset management, maintenance, etc.

Considering that IIoT is effectively IoT in the industrial sector and that the Industrie 4.0 concepts are effectively a subset of IIot, as shown in Fig. 5.1, one needs to identify the difference between IoT and IIoT. Although the basic concepts are the same, i.e., interconnected smart devices that enable remote sensing, data collection, processing, monitoring, and control, the parameters that identify the IIoT subset of IoT are the strong requirements for continuous operation and safety as well as the operational technology employed in the industrial sector. As an example, one can consider the difference between a consumer service such as a health monitoring application on a smart watch and an industrial service such as the monitoring of a steam pump. Although both applications collect real-time data, e.g., steps or body temperature in the health application case and pressure or steam volume in the steam pump case, transmit the data, identify events, and provide feedback or commands to operators/consumers and subsystems, clearly, continuous operation and safety place stricter requirements in the steam pump case, where the potential effect of a failure is significantly more catastrophic and may lead to costly operation down time and even human injuries or loss of life.

These characteristics of the industrial sector – technology and requirements – lead to specialized, demanding solutions for technology and services, justifying the focus of the industrial sector on a specialized IoT concept. This has resulted to the strong interest of the industrial sector in the development of specialized concepts, from strategy to application and technology. The conventional business development models that include numerous interdependencies between stakeholders, from supply chains to service promotion, lead also to a strong need for interoperable solutions at many levels, from the device level to services. Thus, there is need for coordinated activities in the evolution to IIoT, which is addressed by consortia, such as the Industrial Internet Consortium [IIC14] that provides significant leadership in this emerging field.

The General Electric company introduced the term *Industrial Internet* in 2012, as a leader of the Industrial Internet of Things, identifying also the technologies of machine-to-machine communication, SCADA, HMI, industrial data analytics, and

Fig. 5.1 IoT, IIoT, and
Industrie 4.0 relationship

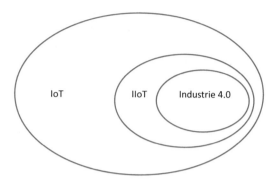

cybersecurity as the main constituents of the IIoT vision [GE17]. Interestingly, they
also calculate the impact of the Industrial Internet to 46% of the global economy,
while in the energy sector they calculate an impact of 100% on energy production
and 44% on energy consumption globally [GE17].

5.4   IIoT Architecture

The development and deployment of IIoT systems and services require the development of architectures that enable efficient and effective operations as well as interoperability considering the anticipated end-to-end services and the large number of
stakeholders involved for devices, cyber-physical systems, communication systems
and networks, service providers, and business developers. Thus, significant effort is
being spent to develop standards and reference architectures that will be accepted
and adopted by the various stakeholders. The International Telecommunication
Union (ITU) has addressed this issue, publishing in 2012 the ITU-T Y.2060 recommendation, which introduces a reference architecture for IoT, in general, including
explicitly applications that fall in the context of IIoT, such as smart grid, intelligent
transportation systems, e-health, etc. [ITU12]. The Industrial Internet Consortium
(IIC) has also been working on a reference architecture for IIoT and currently has
published Version 1.7 of the Industrial Internet Reference Architecture [IIC17].
This architecture is an elaborated reference architecture, significantly more detailed
than the ITU one, addressing all important aspects to all categories of stakeholders.
Taking into account the details of both reference models, one can consider the IIC
model as a specialized evolution of the ITU model, addressing in more details the
important issues of IIoT relatively to the more generic ITU reference model that
encapsulates the requirements for the general IoT.

The ITU effort has expanded the communications' vision to include communication of "anything" to the communication concepts of "any time" and "any place."
Importantly, it includes all expected applications, including industrial ones, specifically mentioning smart grids and intelligent transport systems among others. As
"things," ITU considers physical and virtual objects that are identifiable and able to

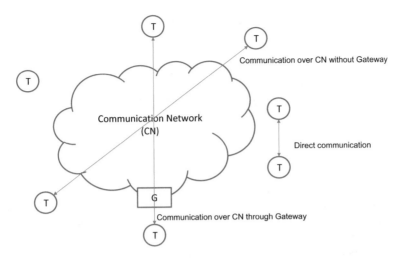

Fig. 5.2 Communication methods for IoT devices

connect to communication networks, while they have related information that is either static or dynamic. Importantly, since communication is a critical part of the whole IoT concept, physical things need to be attached to "devices" that are connected to networks, so that any analog information can be converted to digital and transmitted through the networks. Devices can be simply data-carrying communicating and storing data, data-capturing interacting with the physical objects through reader and writers, sensing and actuating devices, or general-purpose devices with embedded processing and communication resources, such as machines, appliances, and consumer electronic products.

An important issue in the ITU reference model is the communication model among devices. As Fig. 5.2 indicates, the model considers three methods of communication, based on the employment of gateways (G) and the use of the communication network (CN). Devices can communicate without the use of gateways, directly, over local networks, and/or over the communication network, or they can communicate over the communication network exploiting gateways.

The ITU model accommodates fundamental characteristics of IoT that are identified. These fundamental characteristics are interconnectivity, scale, heterogeneity, services for things and the dynamic nature of device information, and connectivity. Interconnectivity is a significant characteristic because "anything" can connect to the global network for any application. As the number of connected devices increases dramatically, scaling becomes a significant parameter that needs to be addressed at all levels of IoT and IIoT; the scaling issue relates not only to communication end points and number of devices but to the size of produced and communicated data as well as their management in terms of storage and processing. The dynamic nature of devices, which turns on and off dynamically or connect and disconnect to networks, will make the landscape more complex and demanding. The open nature of (I)IoT and the large number of stakeholders, in addition to the flexible and long

Fig. 5.3 IoT reference
model by ITU

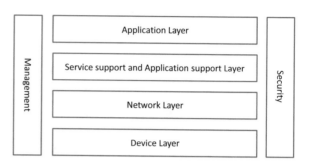

supply chains in conventional service provisioning, leads to the need to accommodate heterogeneous "things," devices, platforms, and services. Services for things also need to be addressed appropriately, not only because of the limited resources of many "things" but also because of the requirements of several services for security and safety, including privacy protection and safe actuation that avoids problems and accidents.

The fundamental characteristics of (I)IoT lead to requirements that need to be met by the reference architecture. The main requirements mentioned by ITU include interoperability, identification-based connectivity, autonomy in networking and services, accommodation of location-based services, security and privacy, as well as capabilities for management of things and services, including plug and play.

Figure 5.3 depicts the ITU IoT reference model, which has been introduced to meet the above requirements. It is a typical layered model with four hierarchical layers, specifically device, network, application and service support and finally application layer, and two vertical layers that are crosscutting the four hierarchical layers, defining management and security functions and properties to all hierarchical layers.

The device layer, the lowest in the hierarchy, includes the functionality of devices and communication gateways. Considering the main interest of ITU in communications, the layer describes communication-centered functionality for the devices: (a) devices that transmit and receive information over the communication network directly, i.e., without using any gateway, (b) devices that communicate information (transmitting and receiving) through gateways, (c) devices that communicate directly without the use of the communication network but being able to communicate over local networks or to form ad hoc networks, and (d) devices that are able to selectively turn on and off functionality in order to save operating power. In regard to gateways, the device layer includes all relevant communication technologies, wired and wireless, such as CAN bus, Wi-Fi, Bluetooth, Zigbee, etc. Importantly, the device layer includes protocol conversion, because devices may implement different protocols, and, thus, needs protocol conversion for interoperability.

The network layer provides encapsulation of device data and related protocol conversion to network layer protocols. The layer includes functionality for the network and transport layers in the OSI protocol reference model. For networking, they include control functionality for network connectivity, mobility, authentication, authorization, and accounting, while for transport they anticipate user traffic transport as well as the transport of control and management information for (I)IoT service and applications.

The service support and application support layer includes both generic and service/application-specific functionality (capabilities) that enable (I)IoT applications and services. Considering the distributed nature of (I)IoT services and applications, there exists generic functionality, such as data processing and storage, as well as specialized functionality, per application and service, since emerging services have different requirements, for example, smart grid operation places different privacy requirements than an intelligent toll management system for transportation services.

Finally, the application layer, the highest hierarchical layer, includes the (I)IoT applications and services.

The management vertical, crosscutting layer includes both generic and application domain-specific functionality. The generic one refers to the typical management for configuration, topology, resource, performance, fault, security, and account management. The application-specific one refers to functions that meet application requirements, such as smart meter monitoring in smart grids.

Analogously to the management layer, the security vertical, crosscutting layer includes both generic and application domain-specific functionality. The generic functionality refers typically to functions related to authorization, authentication, integrity and confidentiality at all layers, privacy at the application layer, secure routing at the network layer, access control at all layers, etc. Application-specific functionality refers to meeting application-specific requirements.

The ITU reference model document presents also a set of business models for IoT, considering the large number of stakeholders in the area and their different interests and goals. Importantly, these business models are developed based on the view of network operators. The business models are based on five main business roles that the stakeholders may have: (a) device provider, (b) network provider, (c) platform provider, (d) application provider, and (e) application customer. As the terms indicate, device providers are the stakeholders that provide devices for (I)IoT, and network providers provide network systems, gateways, and connectivity for the (I)IoT. Platform providers provide the unified, distributed IT platform with well-defined interfaces, over which an application can be served end to end, while application providers are the ones who provide the (I)IoT service over the platform, networks, and devices provided by the corresponding providers. Apparently, the application customer is the user of the (I)IoT application or service.

Based on these five business roles, ITU identifies five business models depending on the number of operators that are involved in an application and their specific roles. Figure 5.4 shows these five models (Models 1–5), presenting the business roles as stacked boxes – analogously to the vertical layer model – and indicating operators with different fill patterns in the boxes; boxes (roles) with the same fill pattern in a stack indicate that the same organization is the operator of these boxes. In Model 1, for example, the same organization has the roles of device, network,

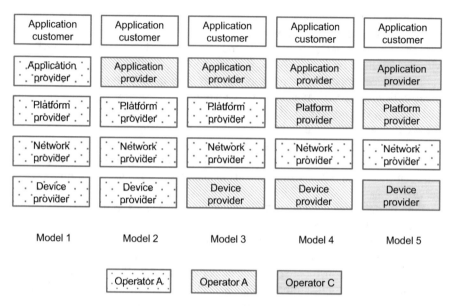

Fig. 5.4 IoT business models identified by ITU

platform, and application provider, while, in Model 2, one stakeholder has the roles of device, network, and platform provider and another one has the role of the application provider.

The Industrial Internet Consortium (IIC) focuses on similar concepts and develops a reference IIoT architecture that has several similarities with the ITU approach and reference model. Clearly, the IIC approach to the architecture development addresses the interests and concerns of all types of stakeholders in an integrated way, originating from use cases and focusing on complete business models and applications at all levels, from devices to IIoT services. IIC follows the approach that different stakeholders who need to make different decisions have architectural viewpoints that are at different levels of abstraction. These viewpoints enable stakeholders to focus on the parameters of interest and develop appropriate architectures that achieve their goals and address the problems they have identified. For this purpose, IIC has identified four different viewpoints: (a) business, (b) usage, (c) functional, and (d) implementation.

The business viewpoint addresses the concerns of business stakeholders, who define and specify IIoT systems and services in their organizations or for customers. These concerns, such as return on investment, cost of maintenance, and similar, are addressed through a model that enables the definition of visions and values which are translated to key objectives and then to high-level specifications of business tasks, named fundamental capabilities. The stakeholders involved include business developers as well as system engineers and product managers.

The usage viewpoint describes how the system is used, implementing the key objectives and the capabilities that have been specified through the business viewpoint. The viewpoint is described with a model that identifies the system and its activities, the involved parties – humans or machines – and their roles, and, finally, tasks, i.e., actions that are executed by parties with a specific role. As tasks are the actions in the system, they are precisely specified and described per role with, so called, functional and implementation maps that specify the exact functions and implementation subsystems that are necessary for a task's complete execution. The stakeholders involved in the usage view include not only the systems engineers and the product managers of the related employed products but all stakeholders that are involved in IIoT system and service specification, including the end users.

The functional viewpoint presents the functional architecture of the IIoT system, describing its components, dependencies, and coordination, meeting the requirements and specifications that have been developed through the usage viewpoint. The stakeholders involved in this viewpoint are system and subsystem developers, product developers, and managers as well as system integrators.

Considering the focus of IIC on IIoT and the increasing adoption of industrial control systems (ICS) within the industries of several sectors and in the operation and management of critical infrastructure, the IIC reference model focuses on its functional architecture of IIoT systems on the integration of ICS with classical information technology (IT) systems in a unified, effective model that meets the requirements of all stakeholders – as specified in the business and usage viewpoints – and enables their effective decisions. The inclusion of ICS and IT in a unified model presents several challenges. Industrial control systems, the systems of Operational Technology (OT), have been developed following a different evolution path from typical IT systems, because of their goals and requirements that typically include continuous operation, safety, and real-time constraints; OT systems have been mostly developed and owned by control and operations engineers, they employ different technologies for processing, communications, and interfaces because they interface directly with the environment through sensors and actuators, and they are managed by their owners independently, since they are typically part of demanding systems and services in terms of dependability, continuous operation, real time, and safety. As a result, their technologies, practices, and standards have evolved independently from the ones for IT. However, the increasing capabilities offered by advanced sensors and actuators, processors, and memories have enabled ICS to execute highly complex operations that have been developed for complex IT systems, such as high-volume data collection and analysis, multivariable modeling and optimization, etc. Importantly, at the same time, the increased capabilities and the increasing complexity of ICS have led them to be more vulnerable to failures and cyber-attacks, leading to additional functional requirements for their correct and efficient operation.

In order to address the integration of IT and OT in a unified model, the IIC approach to the reference architecture divides IIoT systems in five domains, each one grouping the functionality required for a logically distinct high-level operation of the system. These five domains are (a) control, (b) operations, (c) information, (d) application, and (e) business. Figure 5.5, from [IIC17], illustrates the decomposition of the functional representation of an IIoT system into the five domains and

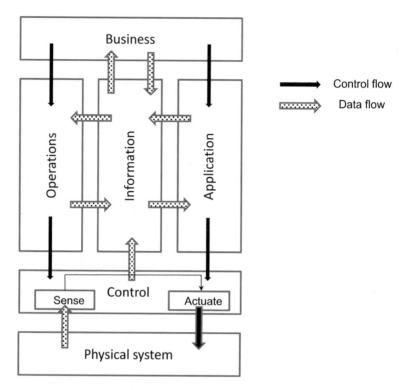

Fig. 5.5 The IIC reference architecture functional domains

shows the data and control flow among the domains, as specified by IIC. The control domain effectively represents the control loop realized by industrial control systems, i.e., it contains the sensors, the logic, and the actuation that constitute a plant implemented by one or more industrial control systems. The operations domain includes the functions that are required for the operation of the industrial control systems in the control domain; the operation includes system monitoring and management as well as optimization for the efficient operation of the systems, especially considering the requirements of several application domains for continuous operation, meeting real-time requirements, and achievement of low-power objectives. The information domain is responsible for collecting data from all domains and analyzing them to enable high-level decisions for the system, e.g., coordinating and optimizing the end-to-end operation of several industrial control systems in the control domain. The application domain includes functionality that is application-dependent and effectively includes the models and operation rules of the application at hand; an important part of this domain is the set of APIs and user interfaces so that other applications or human users can use the application effectively. Finally, the business domain includes systems and functions that enable management and decision making at the business level, e.g., with enterprise resource planning systems (ERP), manufacturing execution systems (MES), etc.

It is important to note that the IIC approach is centered around the concept of a control plant, i.e., it addresses all viewpoints around a control loop that implements a plant. Since control loops can be simple, with one system, or complex with multiple systems typically organized in a hierarchy, the IIC functional domain decomposition can be applied at all levels of a hierarchy. Thus, the decomposition of an IIoT system in the domains does not represent a layered approach as the ITU approach, but rather a logical functional decomposition within a layer or across layers in a hierarchy. Because of this, the IIC reference architecture identifies "crosscutting functions" that are effectively hierarchical (or layered) IT infrastructure functions necessary for the development of a complete IIoT application. These functions include connectivity, distributed data management, analytics, intelligent and resilient control, and any other application function that is necessary for the specific application domain or use case. For example, connectivity has to be implemented in a hierarchical fashion, following standards and practices, interconnecting components within an industrial control system or across several such systems, where each system can be viewed as a collection of functions from all five specified domains. Observing the crosscutting functions mentioned, one can realize that they effectively constitute a layered architecture analogous to the one by ITU. In that respect, one can consider the IIC approach and the ITU approach as complementary, with the IIC reference architecture being a generalization of the ITU one, since it includes crosscutting functions analogous to the ITU layers, while it enables the development of more detailed functional models per layer addressing complete control loops and providing support to all types of stakeholders – from device designers to business developers – for effective decision making.

This analogy and complementarity becomes more apparent with the implementation viewpoint, which addresses the implementation details of the functional viewpoint developed for an IIoT system. The implementation viewpoint includes all the necessary technical and technological details that are necessary for the implementation of a complete IIoT system and its application, including system functionality, technological requirements, communication and network protocols, all types of interfaces, and a mapping of the functional blocks that are specified in the functional viewpoint onto typical implementation architectures, such as the three-tier architecture (where the three tiers are the edge, platform, and enterprise) and the layered databus architecture.

5.5   Basic Technologies

The basic technologies that enable the evolution of IIoT are the sensors, cyberphysical systems, and the related communications and networking technologies that enable their connectivity, among them or to other systems, including enterprise networks. As basic technologies, we designate the ones that are all common to all application domains and use cases.

A fundamental technology for IIoT, and IoT in general, is the technology of RFID (radio-frequency identification) which enables the transmission of a microchip's identification information to a reader over wireless media. It is one of the first technologies that enabled and supported the IoT concept, because RFID technology enabled the automatic identification, monitoring, and operation execution related to RFID-equipped tags. For this reason, RFID technology spread widely since the 1980s in the applications for logistics and supply chain management [Fuq15].

Wireless sensor networks (WSN) constitute another fundamental technology for IIoT, considering their widespread employment in industrial automation and their increasing deployment in critical infrastructures. The solutions for effective WSNs need to address a large number of issues, ranging from communication reliability and real-time requirements to low-power communication due to the deployment of a large number of battery-operated sensors in the field. The significant advances in the area have resulted to a large number of potential solutions and standards for reliable and efficient communication in various environments, e.g., WLAN, Zigbee [Zig], Bluetooth [Blu], 6LoWPAN [Mon07, Hui11, She12], etc. Importantly, they have led to the development of smart (intelligent) sensors, even ones that are autonomous and do not need recharging [Eno].

In addition to the low-level communication protocols that are necessary for connectivity, additional, higher-level protocols are necessary to support distributed computing operations and IIoT applications. Such protocols include service discovery, e.g., multicast DNS (mDNS) [Che13], as well as application protocols that are suitable for the various IIoT application domains such as Constrained Application Protocol (CoAP) [She14], Message Queue Telemetry Transport (MQTT) [Mqt16], and Advanced Message Queuing Protocol (AMQP) [Amq14].

5.6 Applications and Challenges

IIoT applications span a wide range of IoT application domains. Operational technology (OT) systems have become the basic computation platform for the operation and management of most critical infrastructure. The high processing and storage capacity of PLCs, their ability to manage real-time applications with high availability, and their easy management by available SCADA systems have made them quite popular as building blocks of large infrastructures beyond the manufacturing floor, for which they were originally introduced. Today, a large portion of infrastructure is based on industrial control systems (ICS) and makes this critical infrastructure a potential provider of IIoT services and user of IIoT technology. The energy sector is probably the most demanding one on the use of ICS, since the production and processing of energy is part of a country's heavy industry and thus, naturally, includes large ICS platforms. In addition, ICS are used heavily in power distribution networks, such as the electricity network. Considering the emerging smart grids that provide monitoring devices, i.e., PLC-like systems, to customers, it becomes apparent that ICS are the main computing infrastructure in power systems end to end, from production to consumption.

A large number of distribution networks follows this ICS-based model of operation, including water distribution and management networks and water processing sites. Importantly, oil and gas distribution networks use this technology managing pipelines and storage tanks as well as the overall network's operation. Transportation also presents a significant area of application, where ICS and other cyber-physical systems are used for traffic management, i.e., operation and management of traffic lights, for toll payment, etc.

All these application areas of IIoT will require additional deployment and adoption of components, especially cyber-physical and ICS in particular, in order to provide the envisioned services at a large enough scale to improve the lives of citizens significantly. The IIoT revolution is still at its beginning. In this evolution process, the sectors that currently depend on ICS technology will be the forerunners of IIoT technology and will provide the leadership in IIoT development. Currently, the power sector and especially the electricity production, distribution, and consumption processes are the most mature ones, having large bases of ICS at most stages of the service provisioning infrastructure. Despite its maturity, the sector presents quite challenging problems for its next generations. We present some of these challenges here, as a sample illustration of the continuous challenges that need to be resolved in the path to IIoT. Analogous problems exist in other IIoT application areas as well, but the scope of our presentation is to illustrate directions and not to enumerate problems in all application domains.

Stability and continuous operation of the power production and distribution systems constitutes a critical requirement for the development of modern economies. Monitoring the state of the power grid system is a challenging process that is feasible through advanced techniques for fault diagnosis and identification. In this direction and considering the advances in smart grids, we need more advanced techniques for fault detection and isolation in environments with distributed, interconnected power generators. Detection methods based on $\chi2$ distribution statistics enable one to identify, with a high degree of confidence, whether the grid operates well or if there is a fault; furthermore, fault localization and isolation can be achieved by applying such techniques in segments of the grid. Conventional methods for distributed fault diagnosis are limited though, because they do not address the nonlinear dynamics of the grid's behavior, using either algebraic methods that do not address the dynamics or sets of linear differential equations that do not address the nonlinear characteristics. Currently, there is significant effort to develop methods for distributed fault diagnosis taking into account the nonlinear dynamics, focusing on nonlinear modeling, nonlinear state estimation, nonparametric state estimation, development for statistical criteria for fault diagnosis and isolation as well as observability, and diagnosis with distributed sensor networks in the power grid [Rig11, Rig13, Rig15, Rig17].

Power optimization of large consumers, such as large organizations or buildings like hospitals, etc., is a significant challenge which can be addressed by IIoT. Data collection and preparation for processing are critical to the implementation of innovative power management and control. Actually, data preparation is emerging as a critical, time-consuming process especially in heterogeneous environments, requiring the adoption of new and innovative methods for data cleaning, accounting, grouping, and conversion, so that data are presented to processing in a homogeneous fashion. A promising direction to the optimization of power consumption is the identification of patterns in consumption, based on the collected data. Pattern recognition methods play an important role here in two directions, specifically recognition of patterns based on real consumer behavior and development of desired patterns that lead to lower consumptions [Kok09, Hat11, Kou11].

Installation of IoT technologies at a large scale, as in the case of buildings, energy networks, and production lines, requires appropriate processes, mechanisms, and tools. The tools for deployment and configuration for the IoT, and especially IIoT, subsystems constitute a challenge because of the high complexity and heterogeneity of the cyber-physical systems used [Ant16]. The problem becomes more acute when considering the limited resources of wireless embedded systems, the strict requirements for initialization of secure wireless connections, and the requirements for monitoring the parameters that are used for scheduling in real-time wireless networks, such as IEEE 802.15.4e, IETF 6TiSCH6top, and ISA 100.11a [Kou16]. In contrast with small-scale deployments, e.g., in home environments, installation processes at a large scale are error prone, despite their formalization, and lead to installations that have significant costs for reinstallation or reconfiguration when new devices are added or when changes are made, e.g., an office floor reconfiguration. A characteristic example of a formalized, but error-prone, installation method is the "outside-in" installation sequence, where sensors, actuators, and controllers can be installed by technicians before the network, and IT infrastructure in the building is installed. Clearly, it is necessary to develop effective tools for the management of IIoT resources such as wireless sensors and their networks.

IoT technologies, in general, are easily adopted in the industrial and enterprise environments [Bi14], while the addition of wireless cards for the identification of products and materials enables the management of their complete life cycle [CEP10]. Thus, there is a need to integrate these smart and identifiable objects in the industrial enterprise infrastructure and processes. Considering the heterogeneity that characterizes industrial enterprise environments and its layered management, from high-level ERP systems to low-level production management systems, the integration of these devices achieving interoperability is a clear challenge. However, when the challenge is met, the resulting system enables the flexibility of industrial processes and their mapping and distribution on "things" of the IIoT, increasing autonomy within the enterprise.

References

[Amq14]   ISO/IEC 19464:2014 (2014). Information technology: Advanced message queuing protocol (AMQP) v1.0 specification.
[Ant16]   Antonopoulos, C., et al. (2016). Integrated toolset for WSN application planning development commissioning and maintenance: The WSN-DPCM ARTEMIS-JU Project. *MDPI Sensors*.

[Bi14]   Bi, Z., Xu, L. D., & Wang, C. (2014). Internet of Things for enterprise systems of modern manufacturing. *IEEE Transactions on Industrial Informatics, 10*(2), 1537–1546.

[Blu]   Bluetooth specifications. https://www.bluetooth.com/

[CEP10] (2010, March). CERP-IoT vision and challenges for realising the Internet of Things. CERP-IoT – Cluster of European research projects on the Internet of Things.

[Che13] Chesire, S., & Krochmal, M. (2013). Multicast DNS. *IETF RFC, 6762.*

[Eno]   EnOcean Alliance. https://www.enocean-alliance.org

[FoF]   Factories of the Future. European factories of the Future Research Association (EFFRA). http://ec.europa.eu/research/industrial_technologies/factories-of-thefuture_en.html

[Fuq15] Al- Fuqaha, A., et al. (2015). Internet of things: A survey on enabling technologies protocols and applications. *IEEE Communications Surveys & Tutorials, 17*(4), 2347–2376.

[GE17]   GE Digital. Industrial Internet insights from GE Digital. https://www.ge.com/digital/content/industrial-insights-from-ge-digital

[Hat11] Hatziargyriou, N. (2011, September 13–15). Network of the future. *Presentation on behalf of CIGRE TC at the panel session "The electric power system of the future: an international overview". CIGRE international symposium, "The electric power system of the future: Integrating supergrids and microgrids".* Bologna, Italy.

[Hui11] Hui, J., & Thubert, P. (2011, September). Compression format for IPv6 datagrams over IEEE 802.15.4-based networks. IETF RFC 6282.

[IEC16] OSI. Information technology – Security techniques – Information security management systems – Overview and vocabulary. ISO/IEC 27000:2016. http://ww.iso.org

[IIC14] Industrial Internet Consortium. http://www.iiconsortium.org/

[IIC17] IIC. (2017). The industrial Internet of Things volume G1: Reference architecture. IIC:PUB:G1:V1.80:20170131.

[Ind14] Germany Trade and Invest. (2014, July). Industrie 4.0 smart manufacturing for the future.

[ISA16] ISA. (2016, December). The 62443 series of standards – Industrial automation and control systems security. ISA. http://www.isa99.isa.org/Public/Information/The-62443-Series-Overview.pdf

[ITU12] ITU-T. Overview of the Internet of Things. ITU-T SERIES Y: Global information infrastructure Internet protocol aspects and next-generation networks, recommendation Y.20606/2012.

[Kok09] Kok, K., et al. (2009, June 8–11). Smart houses for a smart grid. *20th international conference and exhibition on electricity distribution: Part 1, CIRED 2009,* Prague.

[Kou11] Kourtis, G., Hadjipaschalis, I., & Poullikkas, A. (2011). An overview of load demand and price forecasting methodologies. *International Journal of Energy and Environment, 2,* 123–150.

[Kou16] Koulamas, C., Giannoulis, S., Fournaris, A. (2016). IoT components for secure smart building environments. *Components and services for IoT platforms: Paving the way for IoT standards.* Springer.

[Man13] Manyika, J., et al. (2013, May). Disruptive technologies: Advances that will transform life business and the global economy. McKinsey Global Institute www.mckinsey.com/mgi

[Mic17] Tanner, P. (2017, June 28). Micron benefits from memory market's faster growth rate. Market Realist. http://marketrealist.com/2017/06/micron-benefits-from-memory-markets-faster-growth-rate/

[Mon07] Montenegro, G., Kushalnagar, N., Hui, J., & Culler, D. (2007, September). Transmission of IPv6 packets over IEEE 802.15.4 networks. IETF RFC 4944.

[Mqt16] ISO/IEC 20922:2016 Information technology: Message queuing telemetry transport (MQTT) v3.1.1

[Rig11] Rigatos, G. G. (2011). *Modelling and control for intelligent industrial systems: Adaptive algorithms in robotics and industrial engineering.* Springer.

[Rig13] Rigatos, G. (2013). *Advanced models of neural networks: Nonlinear dynamics and stochasticity in biological neurons.* Springer.

[Rig15]   Rigatos, G. (2015). *Nonlinear control and filtering using differential flatness approaches: Applications to electromechanical systems.* Springer.

[Rig17]   Rigatos, G. (2017). *Intelligent renewable energy systems: Modelling and control.* Springer.

[She12]   Shelby, Z., Chakrabarti, S., Nordmark, E., & Bormann, C. (2012, February). Neighbor discovery optimization for IPv6 over low-power wireless personal area networks (6LoWPANs). IETF RFC 6775.

[She14]   Shelby, Z., Hartke, K., & Bormann, C. (2014, June). The constrained application protocol (CoAP). IETF RFC 7252.

[Zue10]   Zuehlke, D. (2010). SmartFactory—Towards a factory-of-things. *Annual Reviews in Control, 34*(1), 129–138.

[Zig]   ZigBee specifications. http://www.zigbee.org/

Chapter 6
Security and Safety

6.1 Introduction

The Internet of Things (IoT), including the Industrial Internet (IIoT), refers not only to the connectivity of systems and devices but to the related applications and services that provide monitoring and control of complex systems and services. The application domain spans a wide range of industries, from health to industrial control and from transportation to surveillance systems. Its expansion and growth incorporate several technologies and disciplines, such as electronics, embedded networks, hybrid systems, and control. The inclusion of information technology (IT) as well as operational technologies (OT) creates a challenge for the development of systems and services that are technologically interdisciplinary. The resulting challenges to integrate these technologies in new design methodologies for robust and effective IoT systems and services are significant. Currently, even the terminology used by different stakeholders presents challenges and inconsistencies to the common understanding of properties and goals of IoT infrastructure and applications.

Considering the targeted applications and services of IoT, in this chapter, we address security and safety of IoT systems and services with an approach that spans from systems to applications (services or processes) in a unified way, using terminology that originates from computing, networking, and control, since these disciplines constitute the main pillars of IoT technologies in all IoT application domains. This approach is consistent with the reference architectural models of both ITU and the Industrial Internet Consortium, as presented in Chap. 5. For convenience, we address security in this chapter following the ITU model, which divides security mechanisms in two parts, one for generic security and one that is application dependent; we use the terms application dependent and process dependent interchangeably.

IoT applications, in general, collect data through sensing devices, process this data, and take actions that range from sending notification and raising alarms to taking actions through actuators on physical systems. A simple generic model for this

© Springer International Publishing AG 2018
D. Serpanos, M. Wolf, *Internet-of-Things (IoT) Systems*,
https://doi.org/10.1007/978-3-319-69715-4_6

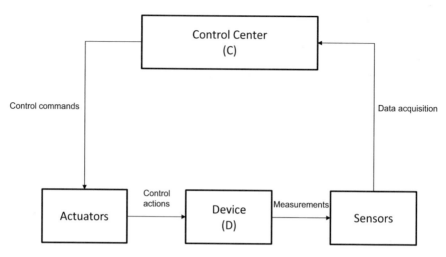

Fig. 6.1 Control loop

operation is the model of the control loop that is used across many application domains and is depicted in Fig. 6.1. In this model, a device D is controlled by a control center C. Measurements of the parameters of interest are collected from D through sensors and delivered to C which makes the necessary calculations and takes the necessary decisions and actions for the application; if the application requires automatic actions, C sends the necessary commands to actuators that control D. The model is generic and covers application across domains ranging from health to transportation and from aerospace to manufacturing. In a health application, for example, sensors measure patient parameters, such as temperature and glucose levels, and send them to a monitoring program – analogous to the control center – and decisions are made depending on the application; a message may be sent to attract a patient's or a doctor's attention, or an insulin pump may be opened to administer more insulin. In a manufacturing floor, sensors may detect the arrival of a component and send the data to a control center which, in turn, sends the appropriate commands to the machine that will process the component accordingly.

The control loop model shown in Fig. 6.1 is implemented on a computational platform that has a different structure from the one indicated in the control model. Figure 6.2 shows a typical hierarchical computational structure for industrial systems, an important class of IoT systems, showing how the computing systems, networks, sensors, and actuators are typically used to implement the operational computing infrastructure of the control loop. Sensors and actuators are attached to the controlled device (D in Fig. 6.1), programmable logic controllers (PLCs) implement simple controls – one per PLC typically – and the supervisory control and data acquisition (SCADA) system implements the control loop for the complete process, also denoted as plant. The PLCs in the structure are simple industrial computers, and their number differs according to the application. In a smart grid, for example, different PLCs may take actions locally per transformer, while SCADA controls the

Fig. 6.2 Hierarchical computing structure for control loop

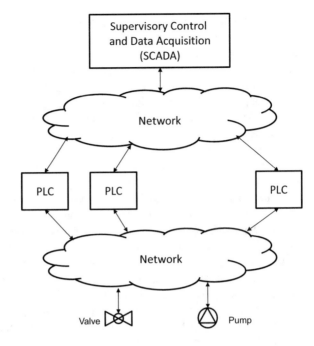

complete smart grid; in a water management system, a different PLC may control each pump, while SCADA controls the water system of an industrial site.

In this environment, there are several properties we want to achieve. From the control point of view, these properties are typically safety properties. For example, we want to avoid overloading of a smart grid, to avoid the overflow of a fluid tank, or to avoid overdose of a pharmaceutical substance that is automatically administered to a patient. These properties can be violated because of several reasons. A programmer may have introduced a bug in the program, the requirements of the system may have missed a condition that should had be taken into consideration, the middleware of the system may give the wrong priorities to control processes, or, simply, a malicious party may attack the system and cause it to take the wrong actions.

The safety requirements for applications are typically expressed as requirements on the control loop which implements applications. These expressions are based on assumptions about properties of the infrastructure on which the application is implemented. For example, an HVAC control system assumes that the temperature measurements that are input to the system are correct within some approximation. This implies that the safety properties are based on assumptions for data integrity that need to be satisfied by the infrastructure. In general, safety requirements include infrastructure security ones, such as integrity, implicitly or explicitly. A typical explicit security property is the protection of personal information in a health management system. Thus, it becomes clear that security is a requirement for safety as well, since data integrity is necessary at least.

IoT technologies involve several stakeholders, including vendors, service providers, regulators, and customers. Although the interests of the independent stakeholders are different and, thus, the security requirements they place on IoT technologies may differ, there is a set of core security requirements that, in general, addresses the requirements of all different categories of the stakeholders. This set of requirements includes (1) confidentiality, (2) integrity, (3) authentication, (4) access control, (5) non-repudiation, (6) dependability, (7) safety, and (8) privacy [Ser13].

Confidentiality is the property that provides protection of data, stored or transmitted, from being disclosed, while integrity enables the confirmation (verification) of the correctness of the related data. Authentication enables the identification of any party involved in a transaction, whether producing, processing, transmitting, or receiving data. Access control ensures service provision to authorized users, while non-repudiation disables participants to transactions to deny actions or their participation. Dependability requires provision of system and service functionality with specific properties such as continuous service even in the presence of errors and failures, meeting specific real-time requirements, etc. Safety is a service and process requirement that warrants service provisioning so that there is no hazard to users. Finally, privacy protects personal information from access by unauthorized actors.

Scientific and engineering methods and techniques to meet these requirements are known, in general, because such requirements have been long addressed in several IT systems in a wide range of application domains. However, meeting the requirements in the IoT and IIoT context with OT characteristics requires new approaches, because of several additional factors. These factors include the models of component failures, the available resources for security provisioning, as well as the profile of attackers, including their potential resources. These factors are strong differentiators in the process of security provisioning in the IoT and IIoT context, for several reasons. First, embedded and CPS systems have already been deployed in significantly larger numbers than the non-embedded (typical IT) systems such as servers, laptops, etc. Second, most of these systems are resource limited in terms of computational, communications, and power resources, and their manufacturers place strong low-cost requirements in order to penetrate large consumer markets. As a result, these systems are deployed in various environments, including hostile ones where malicious users get access to these systems for unspecified lengths of time and with unspecified capabilities to tamper with them. A final reason is the strict requirement for safety in several domains such as automotive, industrial, aeronautics, etc.

These differentiating factors of embedded systems place significant demands on their security, because their large deployment numbers and the diverse operating environments, with many unknown or unanticipated characteristics, lead to a large number of potential attackers with varying capabilities. In addition, many application domains place security requirements that are relevant to safety, dependability, and privacy, as in the case of transportation systems, medical systems, surveillance, etc. The necessity to meet all these requirements on systems with limited resources and low targeted cost leads to highly challenging problems and the need for low cost technologies that achieve the required goals.

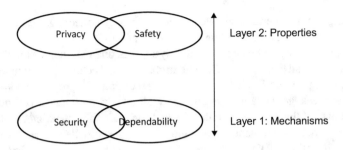

Fig. 6.3 Security property layers

In order to identify the requirements and mechanisms that are required to provide the necessary security properties in the IoT and IIoT context, we follow the layering shown in Fig. 6.3, which has been introduced in [Ser13]. Figure 6.3 defines our view of the relationship between application and process properties, such as safety and privacy, and security and dependability mechanisms which are provided at the system level and are used as primitives to provide the application and process properties.

The depicted layering is based on our approach to differentiate system level properties, such as secure storage, secure communication, tamper resistance, etc., from properties that are required and provided at the application level. In this approach, we consider that (embedded) systems and their interconnections are built to operate resiliently overcoming failures, accidental or malicious, that lead to information loss, leakage, and availability. Dependability mechanisms focus more on the aspects of reliability and availability considering accidental failures, using probabilistic models for the failures, while security mechanisms focus on the provision of alternative properties, e.g., confidentiality, authentication, availability, etc., based on defined malicious attack models. Although some dependability and security properties, such as the availability of information, are common between the two disciplines, others, such as confidentiality or continuous operation, are complementary. In general, dependability is complementary to security, because an attacker can insert faults and failures – analogously to launching attacks on security mechanisms – that the dependability mechanisms cannot recover from. Clearly, the combination of dependability and security mechanisms at the system level provides trusted platforms that are both secure and available under accidents and attacks.

Safety and privacy are often described as security requirements in many application domains, although they are different from the typical security considerations in many ways. Typically, privacy protection and safety are requirements for processes, applications, and services, rather than for generic systems. In our approach, privacy and safety are dependent on security, because they employ security mechanisms for their implementation, such as data integrity and confidentiality. Interestingly, safety and privacy are overlapping, because privacy is a safety issue in some contexts, such as the financial transactions. It is important to note that, as Fig. 6.3 indicates, security and dependability are requirements for privacy and safety. If security mechanisms are lacking, an attacker can violate privacy by easily collecting data or can alter processes and applications, leading to unsafe conditions.

The threat model we consider for IoT systems is one that includes both computational attacks and data attacks. Computational attacks include all malicious actions in a computing system that affect the correct execution of a program and/or lead to information leakage. Data attacks constitute all attacks on input or communicated data. We extend the concept of data attacks to include false data injection attacks, which are malicious interventions that input inappropriate (illegal) data to a system. False data injection (FDI) attacks are an emerging class of attacks to IoT systems, which do not attack the IoT systems themselves but input wrong data to a control system in order to lead it to a wrong decision. In that respect, they are mostly safety attacks. For example, in an HVAC system, a false data injection attack would be to input a higher temperature to the system, instead of the correct measure, in order to lead it to lower the temperature further. Clearly, this type of attacks can lead to hazardous conditions that may endanger processes and systems, even human life.

6.2   Systems Security

IoT systems are embedded computing systems that employ architectures analogous to general-purpose ones. A typical structure of an IoT system is shown in Fig. 6.4, where the system contains four main subsystems: (i) processing, (ii) memory, (iii) input/output, and (iv) power. In general, a secure system requires protection as a whole in addition to protection of all its components individually. The specific requirements are placed depending on the operational environment and the expected capabilities of attackers. In a surveillance system, for example, optical sensors (cameras) need to be secured individually, but the whole network needs to operate dynamically in case individual cameras are compromised or destroyed.

The security of stand-alone systems is achieved with several levels of protection that include physical and hardware security as well as trusted computing platforms. Anti-tampering techniques enable different levels of physical protection ranging

Fig. 6.4 Organization of a typical IoT system

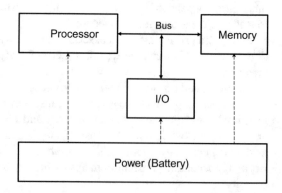

from tamper evidence to tamper response and tamper resistance and are employed accordingly depending on the security requirements of the system and its operational environment. Techniques for tamper evidence simply indicate whether a device has been tampered with. Tamper-response methods combine tamper detection with tamper reaction, where appropriate actions are taken after tamper detection; for example, they destroy stored sensitive data. Tamper resistance methods prevent tampering with devices and protect any sensitive data in the device from attacks.

Anti-tamper technologies have been developed to protect systems after their deployment, so they need to address physical and hardware attacks of attackers with variable capabilities in a wide range of hostile environments, especially for critical applications such as surveillance. They need to combine physical as well as algorithmic mechanisms. Traditional encryption of data, for example, is not a sufficient solution to data protection nowadays, especially in limited-resource systems, where encryption can be overcome with simple attacks. Side-channel attacks have changed the attacks on cryptosystems exploiting physical parameters of the implementations of cryptographic algorithms, such as timing and power consumption, rather than attacking the algorithms themselves [Koc96, Koc99, Qui01] or introduce faults during cryptographic computations [Bar06, Joy09].

Complex hardware systems such as processors and micro-controllers are susceptible to physical and hardware attacks similarly to dedicated circuits, such as cryptographic circuits [And96, Bly93]. Defenses against such attacks require dedicated hardware, specialized design techniques, or even new architectural concepts. For example, a sensitive program can be protected from attacks by storing it in a special design of execute-only memory that allows instructions stored in memory to be executed only and does not allow any other manipulation [Lie00]. Encrypted buses protect data from leakage during data transfers between a processor and its memory [Bes81, Kuh97]. Decay caches can protect from side-channel attacks avoiding cache information leakage [Ker08].

Anti-tampering techniques protect against attacks after system deployment. New business environments can drive embedded systems insecure by planting hardware Trojans during the design and manufacture phase [Jin10].

Embedded and cyber-physical systems, in general, are widespread and have attracted a large range of attacks [Rav04]. Defense against them requires a combination of software and hardware techniques, in order to cover all potential attacks. This is especially important in emerging cyber-physical and IoT systems, which include operating systems or specialized middleware. More complex programmable systems require adoption of such methods as secure booting [Arb97] to establish system integrity, process isolation, and process level attestation techniques [Mic11] to protect running processes as well as techniques for context switching, exception handling, inter-process communication, and memory management [Lie03, Gar03]. Overall, the increasing programmability of these systems requires appropriate software security techniques. Software techniques also offer a cost advantage over hardware techniques. Furthermore, the combination of software techniques with trusted computing modules [Pea02] enables the development of trusted computing platforms for applications and services.

6.3   Network Security

Secure communication requires encryption and authorization mechanisms as well as a secure routing method in a network. Traditional encryption schemes, such as AES [AES01], RSA [RSA78], etc., provide a high level of security, as has been proven in general-purpose computing systems, but they are quite demanding in computational and memory resources. Clearly, they are becoming more viable candidates for adoption in environments where embedded systems obtain increased computational resources. However, today, they are still too demanding computationally for most embedded applications and services. Elliptic curve cryptography provides a promising solution to IoT environments, because it requires lower computational resources than algebraic public key cryptography while providing a high level of security [Miller1986]. Importantly, significant effort is spent to develop and standardize appropriate algorithms for cryptographic primitives for IoT environments, taking into account their characteristics. The development of the Secure Hash Algorithm-3 (SHA-3) by NIST is a significant step in this direction, providing a family of hash functions and extendable output functions that are useful for pseudorandom bit generation, key derivation, and digital signatures in IoT environments [Mor15].

Sensor networks are an important class of IoT subnets that need special attention, because they usually form ad hoc sub-networks with large number of nodes that have very limited computational resources. Thus, sensor network protocols often need to satisfy stricter performance requirements than more complex embedded systems. These limitations typically lead sensor networks to implement cryptographic mechanisms at the link layer. In such limited environments, a good encryption strategy is to use mechanisms with different complexity, depending on the value of the communicated information [Zhu03].

Key management is a critical component of secure IoT communication because keys are the base of cryptographic mechanisms. If key management has weaknesses, keys will be compromised (disclosed or leaked) leading to ineffectiveness of any cryptosystem independently of its strength. The use of global communication keys provides a solution, but such keys cannot be predefined in networked systems usually, because the security of the network can be easily compromised. This leads to the necessity to develop and adopt effective methods to generate and distribute keys. There exist such effective methods mainly using temporary global keys and random key distribution. One such method uses temporary global keys and a global permanent key to establish a main key; then, it destroys the global key in order to avoid key leakage, i.e., the main risk with global keys [Per04]. In an alternative approach, one can use random key distribution. In this case, the system uses a large number of keys and performs communication choosing random subsets of keys.

When the key set sizes are chosen appropriately, all network end points of a network can communicate successfully [Cha03].

Networked systems, especially through the Internet, need to ensure that data are being communicated only among authorized users and processes and that these exchanged data are "legal." This is usually achieved through the use of firewalls, which are typically implemented at the network and application layers, in end point systems or in the network infrastructure [Bol95]. IoT systems typically have very well-defined communication needs, and thus, firewalls can be easily configured to allow strictly the limited type of legitimate communication. The decision about where the firewall should be implemented, i.e., at the network or application layer, at the end system, or in the network, depends on the end point system, the network and their available resources, as well as on the network topology. For example, ad hoc networks need protection at the node level, while more centralized systems can rely more on network level protection [Sli02].

Denial-of-service (DoS) and distributed denial-of-service (DDoS) attacks are a significant threat against IoT systems or exploiting IoT systems. DoS attacks overload resources, such as processor, memory, and network, of the targeted system, in order to prevent it from performing its intended functionality or serving its users.

In general, there are two basic types of DoS attacks [Hus03]. The first type of attack exploits vulnerabilities, hardware or software, by sending carefully constructed packets to the target system; the typical goal is to crash the target system. Often, such vulnerabilities are exploited because systems are not patched. This makes IoT systems especially vulnerable to these attacks, because many IoT systems are not configured to update their software automatically and a wide population of users is not sufficiently aware of the risks and actions they need to take to protect their systems accordingly

In the second type of attacks, the distributed denial-of-service (DDoS) ones, a large population of compromised systems create vast amounts of network traffic toward a victim system; this traffic is combined with legitimate traffic as well. The overload of the aggregated arriving traffic at the target system overloads its resources and renders it incapable to serve its legitimate users. The recent incident of the Mirai botnet attack [New16] demonstrated clearly that IoT devices are vulnerable to malware injection and they can be effectively used to launch DDoS attacks; in the Mirai case, they attacked an Internet directory service, causing significant and costly disruptions to Internet connectivity worldwide.

DDoS attacks are difficult to stop because they exploit shared network services that are accessed by all systems connected to a network. The current version of the Internet Protocol (IPv4) allows systems to send IP packets with arbitrary values in the source IP address field, making it difficult to identify sources of offending IP packets in many attacks [Wan07]. Current efforts to defend against DDoS attacks are usually based on intrusion detection and traceback schemes for detection, filtering, and tracing of an attack [Pen07]. Intrusion detection employs signature- and anomaly-based detection techniques [Cab01, Wan02], while packet marking [Bel03, Sav01] and packet logging [Sno02] are used for attack traceback.

6.4   Generic Application Security

Interconnected IoT systems provide the infrastructure for distributed applications and services. Currently, the vast majority of deployed and emerging applications follows the client-server model, where remote devices (clients) are connected to servers or the cloud, in general, to deliver information, such as collected data and alarms. The servers typically collect data, monitor the operation and processes of IoT connected devices, and send to the devices control programs or data, in order to adapt their operation accordingly. For example, a medical device may collect information about a monitored patient, deliver it to a centralized server, and receive from the server tuning information to adapt its operation appropriately, e.g., to change the frequency of collected data or to change an algorithm used in its local data processing. A connected car may send reports and alarms related to engine operation and receive a prompt to execute a more detailed test, in case of an alarm or suspicious data degradation.

Considering the emerging application and service models, it becomes clear that IoT systems are distributed systems that execute coordinated processes, where each process is typically a control loop, i.e., a process that, in general, receives sensor data and transmits actuator commands. When the client-server model is adopted, the IoT devices execute simpler control loops, while servers execute hierarchically higher level operations, when they are not simply collecting data. In the Industrial IoT, this hierarchy is expressed through the programmable logic controllers (PLCs) as the local, simpler, and lower level devices (clients) and the supervisory control and data acquisition (SCADA) system as the centralized, higher level server system that monitors and coordinates the complete supervised process.

Based on this hierarchical application model, we consider two levels of distributed application for security purposes. The first level, generic application security support, is the one that provides generic services to the IoT environment, such as system update and upgrade, while the second one, process, is the one that implements the specific process for the specific IoT system, e.g., health, car, industrial, etc.

Generic application security support includes mechanisms to defend against attacks to distributed denial-of-service, secure upgrading, etc. Distributed denial-of-service solutions exploit mechanisms at the network layer, as described in the previous section, extending them where necessary to include specifics from the application configuration at hand, such as the location of the servers.

Upgrading and patching IoT systems constitute another challenge that requires inclusion of security mechanisms, because upgrading and patching open systems up to security risks. The functionality for upgrading and patching is necessary for many reasons; software bugs of deployed software need to be fixed, and new features may need to be added to an IoT system's functionality. However, the ability to transmit code to an IoT system raises the risk that one may attack the system by inserting malicious code instead of the legitimate, intended code. Thus, security mechanisms need to be included in the upgrading services to warrant the secure and safe upgrading of the IoT systems. There are several approaches to this challenge. One can limit or prevent the ability to upgrade software components that manage critical system resources in highly hostile application environments. Alternatively, in safer environments, strict access control mechanisms can be used to enable upgrades of different software components by different operators. Mobile code transmission may be prevented, while wired code transmission may be allowed when connectivity is in a controlled environment. In general, remote management of systems, especially IoT systems with limited resources, requires a secure architecture that addresses the operational environment as well as the profiles of the potential attackers.

6.5 Application Process Security and Safety

Application processes, such as control processes in an industrial environment, are programs that execute the necessary code to calculate the required outputs and implement the process's actions. For example, in an HVAC system, an application process may take as input a request to increase the temperature of the controlled environment, and, as a result, it will calculate the necessary increase for the extracted hot air temperature and its volume and will control and adjust the related actuators accordingly to achieve the result. In a more complex environment such as a smart grid, an identified need or a request to add power to the network will lead to the calculations for the necessary power, the identification of the appropriate generators to activate and, finally, the control of the appropriate actuators that will add the generators to the grid. Such application processes in the (I)IoT environment have safety requirements, which are typically expressed as properties that need to be met; for example, in the HVAC system, the temperature of the hot air needs to be within a specified temperature range. Clearly, security of the involved computing and network systems is a prerequisite for meeting the safety requirements; a compromise of these subsystems can lead to wrong calculations and, thus, to wrong actions that violate the safety properties that are required to be met.

Provision of security and safety in (I)IoT environments is one of the areas where the interdisciplinary nature of IoT expresses itself: safety requirements are application dependent and are set, in most cases, by the engineering of the controlled systems, while security – a prerequisite of safety – requires methods of computer and network security, since the IoT systems themselves are distributed computing systems effectively. Bringing all safety and security requirements together is a challenge that has motivated a lot of research and development work recently and will require significant effort in the future to lead to effective solutions that are easily deployed in the field.

The most promising integrated approach to safety and security, from a computational perspective, is to view the problem as a verification and monitoring one. Since application processes are implemented with programs and safety properties are set by the application designers, e.g., control engineers, one can view the process of developing the application programs as one where the application designers provide the specifications of the application, including the safety properties, and then the software is developed accordingly to meet these specifications and be secure from vulnerabilities overall. In this fashion, the safety and security problem becomes a verification and monitoring problem: first is the verification of the produced application software, i.e., that it meets the set requirements, and second the monitoring of the execution of the verified program in order to ensure that it is not altered and executes as expected, based on the specification.

This approach is a behavioral approach to safety and security, since it is based on the specification of the application process. In this context, application behavior is defined by the executable specification that is the starting point of the approach, and this is the way the term is used in the remainder of this text.

6.6 Reliable-and-Secure-by-Design IoT Applications

The concept of secure-by-design applications is an extension to the principle of correct-by-construction programs introduced half a century ago [Dij67]. The challenge posed by IoT applications is that IoT systems typically include a cyber-physical subsystem that interacts with the environment. Thus, in contrast to the original concepts developed for behavioral models of programs with discrete and linear characteristics, the models for cyber-physical and IoT systems need to accommodate continuous and nonlinear characteristics. A model of the environment is also necessary but challenging, because there exist uncertain environmental variations that affect the behavior of physical subsystems; furthermore, it is necessary to model the environment at different levels of abstraction.

The development of reliable and secure-by-design applications has attracted the attention of several efforts, which focus on the development of effective programming language environments. Ur/Web [Chl96] is a language that enables development of reliable and secure web applications by design. For security, Ur/Web ensures that the produced application does not have vulnerabilities, such as for code injection attacks and SQL injections, while for reliability it ensures that the application will not crash during generation of web pages, it will not produce dead intra-application links, etc. The language guarantees these reliability and security properties through an enriched type system based on dependent. In this fashion, Ur/Web achieves an important result: it provides a unified web model, where a programmer develops web applications in a single programming language that can be compiled to other web standards. In another effort, the Jeeves language focuses on run-time, enabling enforcement of security policies and guarantees that programs do not violate security properties by design [Yan12]. Analogous efforts have been made to apply these approaches in the cyber-physical systems application domain. The ROSCoq framework [Ana15] employs the Coq proof assistant [Ber04] to model cyber and physical resources of robots through an extended logic of events and then to prove various properties of the model. VeriDrone [Mal16, Cha16], a reasoning framework also developed in Coq, ensures security of cyber-physical system models at different but independent levels, i.e., from high level models to C implementations.

6.7   Run-Time Monitoring

Run-time monitoring systems for security can be classified based on two parameters: (i) the method that describes the behavior, i.e., profile based or model based, and (ii) the method that compares the behaviors, i.e., matching to bad behavior or deviation from good behavior. This leads to a classification with four classes, as shown in Fig. 6.5.

Profile-based approaches monitor parameters of the observed system and build a profile of system operation. Class 1 monitoring systems that detect attacks by matching with bad behavior (Class 1 in the figure) typically use statistical methods and machine learning methods to build profiles of bad behavior and statistical profiles of attacks [Hod04, Val00]. They are more robust than model-based systems (Class 2 systems) because machine learning typically generalizes from the collected data, but they suffer from high false alarm rates, and they do not provide rich information for diagnosis when an alarm is raised. Systems in Class 3, which detect deviations from good behavior, usually build a statistical profile of good behavior and detect deviations from that [Kim04, Lak05].These systems are actually more robust than the ones in Class 1, because they do not depend on any past information of attacks and, thus, they raise alarms when new attacks are launched, because all deviations from good behavior are detected. However, not only do they provide limited diagnosis information, i.e., only that something extraordinary has happened, but they suffer from high false alarm rates, because the deviation may not be malicious or accidental, but it can also be normal but just out of the statistically accepted profile behavior.

Model-based monitoring systems, Class 2 and Class 4 systems, use a model of the behavior of the monitored system. Such systems are popular in highly secure

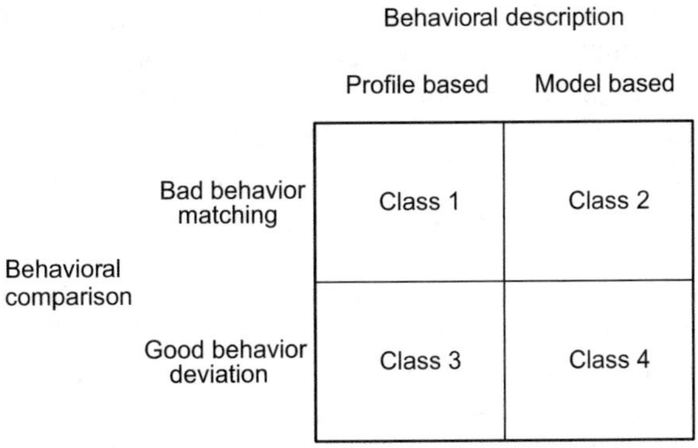

Fig. 6.5 Classification of run-time security monitoring systems

	Behavioral description	
	Profile based	Model based
Bad behavior matching	Class 1	Class 2
Good behavior deviation	Class 3	Class 4

Behavioral comparison

environments, where successful attacks have high cost. Because they use a behavioral model of the observed system, these monitors provide rich diagnostic information when alarms are raised, in contrast to profile-based monitors. Despite this rich information though, Class 2 monitors are limited because they can detect only known attacks; this originates from their bad behavior models which are already known by definition, i.e., the attacks exist [Pax99, Roe99]. Signature-based systems are typical examples in this class. Class 4 monitors detect deviations from a good behavior model [Wat07, Gol07] and thus provide even higher diagnostic information, because there is adequate knowledge of the exact problem, e.g., the exact instruction, that led to a detected deviation. However, the execution overhead of the models of good behavior poses limitations to run-time system performance.

6.8   The ARMET Approach

A promising approach that addresses safety and security in a unified way in IoT systems and cyber-physical systems is the ARMET approach [Kha17]. ARMET is based on three basic concepts: (i) we can build secure-by-design systems, (ii) we can monitor these systems at run-time for correct operation to detect attacks or failures, and (iii) when there is a failure or an attack, we can have plans to recover, depending on the problem and how much information we have about it. ARMET has been developed focusing on industrial control systems, but it is applicable to other IoT systems as well, since their software complexity is comparable to that of industrial control systems.

With the ARMET approach, an IoT application is developed starting from an executable specification which is provably consistent with the safety properties set for the application. From this executable specification, the application code is derived. Given the executable application specification and the application code for the target system, ARMET monitors the behavior of an application while it executes, by comparing its observed behavior to the expected behavior based on the application's specification; to achieve this, a middleware executes the executable specification in parallel with the application execution on the IoT system and calculates predictions of the application's behavior. Figure 6.6 shows the structure of the ARMET middleware system, which is composed of several components: (i) the run-time security monitor, (ii) the diagnosis module, (iii) the recovery module, (iv) the trust model, (v) the adaptive method selection module, and (vi) the backup module. The run-time security monitor is a critical component of the middleware, which takes as input the executable specification of the application and the state of the system that executes the code of the application. The monitor observes the behavior of the application execution, and, in parallel, it predicts the state of the application execution by executing its specification; the specification execution defines the expected "good behavior" of the application and, optionally, known "bad behavior" of the application that includes known attacks. Comparing the predictions with the observations, the monitor can detect deviations that indicate a failure of the

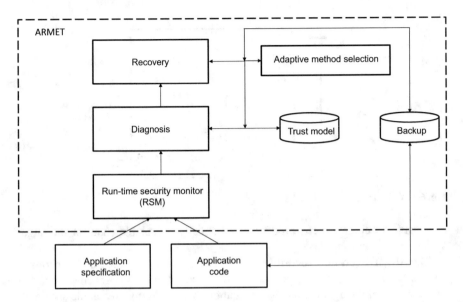

Fig. 6.6 The ARMET system

application or an attack. When such a detection is made, ARMET proceeds to a stage of diagnosis, in order to identify the failure or attack based on a trust model that it includes. After the diagnosis phase is concluded, all available information is used by the recovery module. Based on the diagnostic information, the recovery module chooses an appropriate adaptive method for recovery and enables the system to recover, taking into consideration previous states, as stored by the backup module. It is important to note that the system will operate under all scenarios of failures and attacks, even unknown ones, i.e., failures and attacks that have not been anticipated and are not included in the trust model. In a worst-case scenario, when no useful information is provided by the diagnosis module, the system will recover by returning to a previous clean state. Furthermore, the approach is based on one assumption: the executable application specification is executed in a safe environment and cannot be attacked, i.e., its predictions are always correct; although this may seem as a strong assumption, conventional trusted platforms enable the development of a low-cost IoT platform that meets this requirement and makes the assumption realistic, such as Intel SGX [Cos16] and ARM TrustZone [ARM05].

The process to develop executable specifications and prove its properties is a typical program verification process that can be implemented with various existing tools that enable automated or semiautomated proofs. In the case of ARMET [Kha17], the process is based on Fiat [Del15] and employs deductive synthesis to develop reliable-and-secure-by-design industrial control applications through interactive stepwise refinement of declarative specifications; cyber and physical resources are included as first class models, and nonfunctional properties, such as security and performance, are modeled integrated with functional properties. Apparently, the

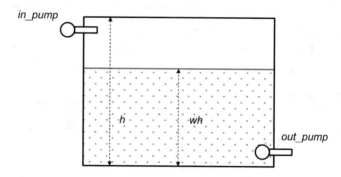

Fig. 6.7 Water tank

executable specification can produce automatically executable code for the target system, IoT or industrial.

The ARMET run-time security monitor (RSM) successfully identifies inconsistencies between predictions, produced by the execution of the specification, and observations of the application code execution, because of its executable specification language [Kha15]; importantly, the predictions are generated automatically. The run-time security monitor (RSM) is the first one to be formally proven as sound and complete [Kha15]; the proof means that the monitor is also free of false alarms (detections), an important, desirable property in practical systems, where false alarms lead to lost resources that are used to explore the false alarms. Importantly, ARMET's specification language allows the specification of faulty behaviors as well as attack plans, which can be used by the monitoring system for threat detection.

The ARMET approach is based on the concept that a system can be specified with an executable specification. Based on an appropriate functional specification for a system, one can express the safety and security properties that the system should meet as conditions of the specification and include them in the specification as well. As an example, let us consider the case of a water tank which has a height h, as shown in Fig. 6.7, and two pumps that are controlled, one for filling the tank with water, denoted *in_pump*, and one, *out_pump*, for draining the water out; each of the two pumps has only two possible states, i.e., open or closed. Furthermore, we assume that there is a sensor that measures the water height, denoted *wh*, in the tank.

We want to have a water management system, where a user issues commands to pour water or drain water from the tank. For simplicity, we consider that a user can perform three actions, FILL, DRAIN, or NOTHING, and that the system operates in cycles, synchronously with a clock. So, during every cycle (clock tick), one action can be performed. A FILL action implies that *in_pump* opens, *out_pump* closes, and for this one time unit water is poured in the tank. A DRAIN action means that *in_pump* closes, *out_pump* opens, and for this one time unit water drains out of the tank. When the action is NOTHING, then both pumps are closed and the state of the tank remains the same. In an environment like this, an obvious safety property is that we do not want the tank to overflow under any conditions.

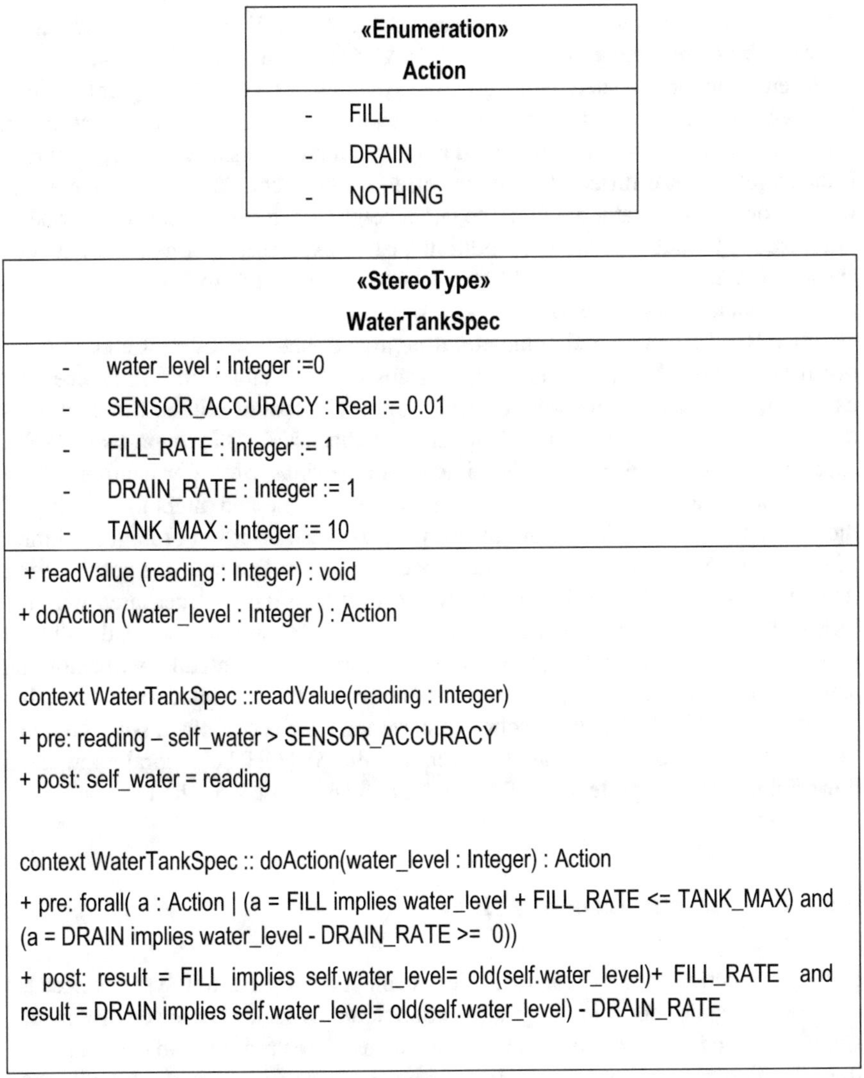

«Enumeration»
Action
-    FILL
-    DRAIN
-    NOTHING

«StereoType»
WaterTankSpec

-    water_level : Integer :=0
-    SENSOR_ACCURACY : Real := 0.01
-    FILL_RATE : Integer := 1
-    DRAIN_RATE : Integer := 1
-    TANK_MAX : Integer := 10

+ readValue (reading : Integer) : void

+ doAction (water_level : Integer) : Action

context WaterTankSpec ::readValue(reading : Integer)

+ pre: reading – self_water > SENSOR_ACCURACY

+ post: self_water = reading

context WaterTankSpec :: doAction(water_level : Integer) : Action

+ pre: forall(a : Action | (a = FILL implies water_level + FILL_RATE <= TANK_MAX) and (a = DRAIN implies water_level - DRAIN_RATE >= 0))

+ post: result = FILL implies self.water_level= old(self.water_level)+ FILL_RATE   and result = DRAIN implies self.water_level= old(self.water_level) - DRAIN_RATE

Fig. 6.8 Water tank control executable specification

Figure 6.8 shows one executable specification, written in UML, which implements the three defined actions, assuming that each action FILL or DRAIN has as a parameter an integer value for the variable *water_level*, which specifies the target height of the water that the user wants to obtain; furthermore, the specification ensures that the water tank never overflows. In the specification, the three actions are defined in enumeration: SENSOR_ACCURACY defines the measurement accuracy of the reading sensor for the water level in the tank, FILL_RATE is the

incoming water rate through *in_pump*, and DRAIN_RATE is the rate of the outgo-
ing water when *out_pump* opens. TANK_MAX is the height *h* of the tank.

When an action is issued by the user, the system first takes a reading of the water
level with the sensor, as specified in *readValue*, and identifies whether the target
water height differs from the measured height within the sensor's accuracy bounds.
If the target height is different, then the corresponding action is performed, pouring
water in or draining water out until the target height is achieved. The safety property
is enforced, because of the precondition that is expressed in *doAction()*, which
ensures that a FILL action is performed when its result leads to a water height that
is less or equal to TANK_MAX.

Since RSM is sound and complete, it is proved that it will detect all computational attacks on the application. This means that any attack that influences the
execution of the application and leads to wrong calculations will be detected. This
has been confirmed with several computational attacks [Kha17]. Importantly, RSM
captures a wide range of false data injection attacks as well. For example, if an
attacker wants to overflow the water tank of the example and alters the reading of
the sensor to a lower value – with the purpose to cause insertion of larger volumes
of water – RSM will identify the attack, because the execution of the specification
will calculate a different value for the water level than the one measured with the
sensor. The difference between the expected water level and the one read will lead
to a detection of the deviation; it will raise an alarm and, eventually, will cause the
action to be stopped. Although there exist complex false data injection attacks that
are not detected by RSM, its detection of common attacks combined with the proof
that it detects all computational attacks makes the ARMET behavioral approach a
powerful tool for the protection of processes and applications in the IoT space.

6.9 Privacy and Dependability

Privacy protection is one of the most significant challenges in IoT systems because
of the legal requirements in many application domains such as home environments,
smart grids, and health systems. There are increasing restrictions and constraints on
the collection, storage, and processing of personal information involved in all applications, including IoT. Privacy protection solutions may need to integrate a range of
methods and techniques, such as time-limited storage of sensitive information,
access control systems to enable access only for authorized personnel, accounting
systems to enable auditing, etc. The burden to comply with the required policies and
laws is further increased by the increasing amount of information considered as
personal or private, which leads to a need for adaptive and scalable solutions that
accommodate new policies as the relevant legal requirements emerge [Mul06]. The
ARMET approach provides a powerful solution to the problem of privacy protection, when privacy protection is viewed as a safety property. Privacy protection
originates from legal requirements that can be expressed as conditions in an information system, i.e., they can be expressed as preconditions, postconditions, or

invariants in a program; for example, a function that is used by unclassified users can be restricted from accessing specific variables that are available only to highly classified ones. With this view, privacy requirements can be expressed as safety requirements, refined into conditions, and enforced with a run-time monitor, like RSM, which will detect all attempts to violate the defined conditions. Importantly, the programmability of the conditions enables dynamic adjustment of the run-time monitors as new conditions are established by emerging legal frameworks.

Interestingly, the behavioral approach to security and safety provides a promising solution to the challenge of combining dependability and security in the same framework [Ser08]. Dependable systems have been developed for a long time with well-understood methodologies, but they are based on fault models that consider faults and errors accidental [Sie82]. In the case of security attacks though, malicious attackers insert faults on purpose, and the models of these faults are fundamentally different from the accidental ones. The behavioral approach to security considers only the attack model, e.g., computational or false data injection, and is not influenced by its origin – accident or on purpose. Thus, it detects accidental faults and malicious attacks with the same method and in the same way. Attribution of the fault is made in ARMET, for example, only after detection and based on the available information and the trust model used. Independently of the attribution though, the behavioral approach will detect a problem, providing a unified approach to security and dependability.

References

[AES01] NIST. (2001). Advanced Encryption Standard. FIPS Publication 197, November 26, 2001.

[Ana15] Anand, A., & Knepper, R. (2015). ROSCoq: Robots powered by constructive reals. In Proceedings of the 2015 *International Conference on Interactive Theorem Proving* (pp. 34–50). Springer LNCS-9236.

[And96] Anderson, R., & Kuhn, M. (1996). Tamper resistance: A cautionary note. In *Proceedings of the 2nd Workshop on Electronic Commerce*, USENIX Association, Berkeley, CA, 1996, pp. 1–11.

[Arb97] Arbaugh, W., Farber, D., & Smith, J. (1997). A secure and reliable bootstrap architecture. In *Proceedings of the IEEE Symposium on Security and Privacy*, 1997, pp. 65–71.

[ARM05] ARM Security Technology. (2005). Building a Secure System using TrustZone Technology. ARM white paper, Document PRD29-GENC-009492C, 2005. http://infocenter. arm.com/help/topic/com.arm.doc.prd29-genc-009492c/PRD29-GENC-009492C_trustzone_ security_whitepaper.pdf

[Bar06] Bar-El, H., Choukri, H., Naccache, D., Tunstall, M., & Whelan, C. (2006). The sorcerer's apprentice guide to fault attacks. *Proceedings of the IEEE, 94*(2), 370–382.

[Bel03] Belenky, A., & Ansari, N. (2003). IP traceback with deterministic packet marking. *IEEE Communications Letters, 7*(40), 162–164.

[Ber04] Bertot, Y., & Castran, P. (2004). *Interactive theorem proving and program development-Coq'Art: The calculus of inductive constructions*. Berlin Heidelberg: Springer.

[Bes81] Best, R. (1981). *Crypto microprocessor for executing enciphered programs*. US patent 4,278,837, July 1981.

[Bly93]   Blythe, S., Fraboni, B., Lall, S., Ahmed, H., & De Riu, U. (1993). Layout reconstruction of complex silicon chips. *IEEE Journal on Solid-State Circuits, 28*(2), 138–145.

[Bol95]   Bolding, D. (1995). Network security, filters and firewalls. *Crossroads, 2*(1), 8–10.

[Cab01]   Cabrera, J., Lewis, L., Qin, X., Lee, W., Prasanth, R., Ravichandran, B., & Mehra, R. (2001). Proactive detection of distributed denial of service attacks using MIB traffic variables—A feasibility study. In *Proceedings of the IEEE/IFIP International Symposium on Integrated Network Management*, pp. 609–622.

[Cha03]   Chan, H., Perrig, A., & Song, D. (2003). Random key predistribution schemes for sensor networks. In *Proceedings of the IEEE Symposium on Security and Privacy*, pp. 197–213.

[Cha16]   Chan, M., Ricketts, D., Lerner, S., & Malecha, G. (2016). Formal verification of stability properties of cyber-physical systems. In CoqPL'16, Jan 2016.

[Chl96]   Chlipala, A. (2016). Ur/web: A simple model for programming the web. *Communications of the ACM, 59*(8).

[Cos16]   Costan, V., & Devadas, S. (2016). Intel SGX explained. Cryptology ePrint Archive: Report 2016/086, IACR.

[Del15]   Delaware, B., Pit-Claudel, C., Gross, J., & Chlipala, A. (2015). Fiat: Deductive synthesis of abstract data types in a proof assistant. In *Proceedings of the 42nd Annual ACM SIGPLAN-SIGACT Symposium on Principles of Programming Languages (POPL'15)*, Mumbai, India, Jan. 15–17, 2015, pp. 689–700.

[Dij67]   Dijkstra, E. W. (1967). A constructive approach to the problem of program correctness, August 1967, circulated privately.

[Gar03]   Garfinkel, T., Rosenblum, M., & Boneh, D. (2003). Flexible OS support and applications for trusted computing. In *Proceedings of the 9th Conference on Hot Topics in Operating Systems* (Vol. 9, pp. 25–25).

[Gol07]   Goldsby, H. J., Cheng, B. H. C., & Zhang, J. (2008). AMOEBA-RT: Run-Time Verification of Adaptive Software. In *Proceedings of Models in Software Engineering (MODELS 2007)*, Nashville, TN, USA, September 30–October 5, 2007, LNCS-5002, Springer, pp. 212–224.

[Hod04]   Hodge, V., & Austin, J. (2004). A survey of outlier detection methodologies. *Artificial Intelligence Review, 22*(2), 85–126.

[Hus03]   Hussain, A., Heidemann, J., & Papadopoulos, C. (2003). A framework for classifying denial of service attacks. In *Proceedings of the conference on applications, technologies, architectures, and protocols for computer communications* (pp. 99–110). New York: ACM.

[Jin10]   Jin, Y., & Makris, Y. (2010). Hardware Trojans in wireless cryptographic ICs. *IEEE Design and Test, 27*(1), 26–35.

[Joy09]   Joye, M. (2009). Protecting RSA against fault attacks: The embedding method. In *Proceedings of the Workshop on Fault Diagnosis and Tolerance in Cryptography (FDTC)*, pp. 41–45.

[Ker08]   Keramidas, G., Antonopoulos, A., Serpanos, D., & Kaxiras, S. (2008). Nondeterministic caches: A simple and effective defense against side channel attacks. *Design Automation of Embedded Systems, 12*(3), 221–230.

[Kha15]   Khan, M. T., Serpanos, D., & Shrobe, H. (2015). On the formal semantics of the cognitive middleware AWDRAT. Technical Report MIT-CSAIL-TR-2015-007, Computer Science and Artificial Intelligence Laboratory, MIT, USA, March 2015.

[Kha17]   Khan, M. T., Serpanos, D., & Shrobe, H. ARMET: Behavior-Based Secure and Resilient Industrial Control Systems. In *Proceedings of the IEEE*, Preprint. URL: http://ieeexplore.ieee.org/stamp/stamp.jsp?tp=&arnumber=8011473&isnumber=4357935

[Kim04]   Kim, S. S., Reddy, A. L. N., & Vannucci, M. (2004). Detecting traffic anomalies through aggregate analysis of packet header data. In *Proceedings of 3rd International IFIP-TC6 Networking Conference (NETWORKING 2004)*, Athens, Greece, May 9–14, 2004, Springer LNCS-3042, pp. 1047–1059.

[Koc96]   Kocher, P. (1996). Timing attacks on implementations of Diffie-Hellman, RSA, DSS, and other systems. In *Advances in Cryptology – CRYPTO'96*. Springer, pp. 104–113.

[Koc99] Kocher, P., Jaffe, J., & Jun, B. (1999). Differential power analysis. In *Advances in Cryptology-CRYPTO'99*. Springer, pp. 789–789.

[Kuh97] Kuhn, M. (1997). The Trust No1 cryptoprocessor concept. http://www.cl.cam.ac.uk/mgk25/.

[Lak05] Lakhina, A., Crovella, M., & Diot, C. (2005). Mining anomalies using traffic feature distributions. In *Proceeding of the 2005 Conference on Applications, Technologies, Architectures and Protocols for Computer Communications (SIGCOMM 2005)*, Philadelphia, PA, USA, August 22–16, 2005, pp. 217–228.

[Lie03] Lie, D., Thekkath, C., & Horowitz, M. (2003). Implementing an untrusted operating system on trusted hardware. *ACM SIGOPS Operating Systems Review, 37*(5), 178–192.

[Lie00] Lie, D., Thekkath, C., Mitchell, M., Lincoln, P., Boneh, D., Mitchell, J., & Horowitz, M. (2000). Architectural support for copy and tamper resistant software. *ACM SIGPLAN Notices, 35*(11), 168–177.

[Mal16] Malecha, G., Ricketts, D., Alvarez, M. M., & Lerner, S. (2016). Towards foundational verification of cyber-physical systems. In *Proceedings of 2016 Science of Security for Cyber-Physical Systems Workshop (SOSCYPS)*, April 2016, pp. 1–5.

[Mic11] MICROSOFT. (2011). Shared source initiative. http://www.microsoft.com/resources/ngscb/default.mspx

[Mor15] Dworkin, M. J. (2015). SHA-3 Standard: Permutation-based hash and extendable-output functions. Federal Information Processing Standards (NIST FIPS) – 202, August 04, 2015.

[Mul06] Muller, G. (2006). Special issue: Privacy and security in highly dynamic systems-introduction. *Communications of the ACM, 49*(9), 28–31.

[New16] Newman, L. H. (2016). What we know about Friday's massive east coast internet outage. WIRED, October 21, 2016.

[Pax99] Paxson, V. (1999). Bro: A system for detecting network intruders in real-time. *Computer Networks, 31*(23–24), 2435–2463.

[Pea02] Pearson, S. (2002). *Trusted computing platforms: TCPA technology in context*. USA: Prentice Hall.

[Pen07] Peng, T., Leckie, C., & Ramamohana-Rao, K. (2007). Survey of network-based defense mechanisms countering the DoS and DDoS problems. *ACM Computing Surveys, 39*(1), Article 3.

[Per04] Perrig, A., Stankovic, J., & Wagner, D. (2004). Security in wireless sensor networks. *Communications of the ACM, 47*(6), 53–57.

[Qui01] Quisquater, J. J., & Samyde, D. (2001). Electromagnetic analysis (EMA): Measures and counter-measures for smart cards. In *Proceedings of the International Conference on Research in Smart Cards: Smart Card Programming and Security*, Springer LNCS-2140, pp. 200–210.

[Rav04] Ravi, S., Raghunathan, A., Kocher, P., & Hattangady, S. (2004). Security in embedded systems: Design challenges. *ACM Transactions on Embedded Computing Systems, 3*(3), 461–491.

[Roe99] Roesch, M. (1999). Snort – lightweight intrusion detection for networks. In *Proceedings of the 13th USENIX Conference on System Administration (LISA '99)*, pp. 229–238.

[RSA78] Rivest, R. L., Shamir, A., & Adleman, L. (Feb. 1978). A method for obtaining digital signatures and public-key cryptosystems. *Communications of the ACM, 21*(2), 120–126.

[Sav01] Savage, S., Wetherall, D., Karlin, A., & Anderson, T. (2001). Network support for IP traceback. *IEEE/ACM Transactions on Networking, 9*(3), 226–237.

[Ser08] Serpanos, D., & Henkel, J. (2008). Dependability and security will change embedded computing. *Computer, 41*(1), 103–105.

[Ser13] Serpanos, D. N., & Voyiatzis, A. G. (2013). Security challenges in embedded systems. *ACM Transactions on Embedded Computing Systems, 12*(1s), Article 66.

[Sie82] Siewiorek, D., & Swarz, R. (1982). *The theory and practice of reliable system design*. Bedford: Digital Press.

[Sli02]   Slijepcevic, S., Potkonjak, M., Tsiatsis, V., Zimbeck, S., & Srivastava, M. (2002). On communication security in wireless ad-hoc sensor networks. In *Proceedings of the 11th IEEE International Workshop on Enabling Technologies*, pp. 139–144.

[Sno02]   Snoeren, A., Partridge, C., Sanchez, L., Jones, C., Tchakountio, F., Schwartz, B., Kent, S., & Strayer, W. (2002). Single-packet IP traceback. *IEEE/ACM Transactions on Networking, 10*(6), 721–734.

[Val00]   Valdes, A., & Skinner, K. (2000). Adaptive, model-based monitoring for Cyber Attack Detection. In *Proceedings of the 3rd International Workshop on Recent Advances in Intrusion Detection (RAID 2000)*, Toulouse, France, October 2–4, 2000, Springer, pp. 80–93.

[Wan07]   Wang, H., Jin, C., & Shin, K. (2007). Defense against spoofed IP traffic using hop-count filtering. *IEEE/ACM Transactions on Networking, 15*(1), 40–53.

[Wan02]   Wang, H., Zhang, D., & Shin, K. (2002). Detecting SYN flooding attacks. In *Proceedings of the 21st Annual Joint Conference of the IEEE Computer and Communications Societies (INFOCOM'02)*, pp. 1530–1539.

[Wat07]   Watterson, C., & Heffernan, D. (2007). Runtime verification and monitoring of embedded systems. *Software, IET, 1*(5), 172–179.

[Yan12]   Yang, J., Yessenov, K., & Solar-Lezama, A. (2012). A language for automatically enforcing privacy policies. In *Proceedings of the 39th ACM Symposium on Principles of Programming Languages (POPL 2012)*, Philadelphia, PA, USA, January 25–27, 2012, pp. 85–96.

[Zhu03]   Zhu, S., Setia, S., & Jajodia, S. (2003). LEAP: Efficient security mechanisms for large-scale distributed sensor networks. In *Proceedings of the 10th ACM Conference on Computer and Communications Security*, pp. 62–72.

Chapter 7
Security Testing IoT Systems

7.1 Introduction

Systems need to be evaluated for conformance to specifications and requirements, including security, and IoT systems are no exception. Verification and validation techniques are one option to ensure that systems are built according to specifications and requirements, but their use is limited for two main reasons: (i) the complexity of these processes is growing exponentially with the size of the checked system, and (ii) the current business models that include long supply chains with different providers and developers of system components do not enable a unified description of designs and implementations that can be checked as a whole. In the case of IoT systems specifically, the size of most systems is not prohibitive for formal verification and validation methods; however, the lack of a complete implementation with the same tools and models leads to fragmented application of verification techniques to components. Testing constitutes an important and necessary phase in system development, which complements all other approaches and enables the evaluation of integrated systems. Thus, testing is an integral part of the systems development cycle with the purpose to evaluate system correctness, performance, and security at least. Importantly, testing is a method used by customers and certification authorities to evaluate conformance of systems to standards and to provide certifications at the device, system, and product level.

The wide deployment of consumer electronics devices has brought significant attention to testing and its methodologies not only for accepting devices by consumers but also for security, since attackers exploit testing methodologies to identify vulnerabilities and exploit them for their purposes. This is especially important to IoT systems which typically have a cyberphysical component. Identification of vulnerabilities in IoT systems and their exploitation may compromise their safety properties and lead to significant operational problems that result to monetary losses, operation disruption, and even loss of life.

© Springer International Publishing AG 2018
D. Serpanos, M. Wolf, *Internet-of-Things (IoT) Systems*,
https://doi.org/10.1007/978-3-319-69715-4_7

Successful testing of IoT systems is critical considering that many of them have strong requirements that are crucial to their operation, such as meeting real-time constraints, satisfying specific safety properties, and continuing operation even under strained conditions. Furthermore, IoT systems include a communication component, which constitutes a testing challenge because the specifications of communication protocols often have undefined parameters that lead to differing implementations by different vendors; this is the reason why interoperability in communication systems is an important challenge. The criticality of IoT testing, especially for security, becomes more apparent when considering industrial IoT systems, which are extensively used in critical infrastructures nowadays, such as energy networks, water management systems, etc. Successful testing not only confirms the expected operations but takes away from attackers the tools to cause malfunctions and disruptions; in the emerging environment, even crashing an application or an operation may be more catastrophic than hijacking them.

Hardware and software testing are technological areas with significant effort in the market and in academia for decades. A large number of methodologies and tools have been developed, but software testing has been a significantly harder problem than hardware testing because of several differentiating characteristics software has, such as evolution through added features and functionality, fault models and lack of re-use. Considering that most IoT systems are built using off-the-shelf hardware components and computing subsystems, we address software testing for security in this chapter, and, more specifically, we focus on the testing of their communication protocol implementations, since it is the point of entry to systems and a common target of attackers. We present fuzz testing, the most common testing approach for security, which requires no information about the internal structure of the system that is tested. As industrial IoT systems constitute an attractive target for attackers that exploit testing techniques, we use as an example the Modbus protocol and describe fuzz testing techniques for its implementations, which give successful results for existing protocol implementations in the field.

7.2   Fuzz Testing for Security

Vulnerabilities in network systems and applications are identified and disseminated publicly [Nis, Sfo, Str]. The cost of fixing these vulnerabilities can be high, while their exploitation may have quite costly consequences. As a result, there is strong research and development effort to reduce such vulnerabilities.

Static analysis of source code is one approach that does not require program execution but is limited because it does not detect vulnerabilities that are activated by dynamic instruction sequences, during program execution, e.g., dependent on subroutine calls [Che07, Vie00]. Also, these methods present a high false-positive rate leading to significant overhead for the evaluation of the results. Alternatively, dynamic analysis methods intervene in program execution. StackGuard, for example, expands a C compiler and produces executable code that identifies potential execution faults – examining addresses, for example – without changing the functionality of the original programs [Cow98]. TaintCheck applies taint analysis for automatic vulnerability analysis without need for the source code [Cla07, New04]. Dynamic analysis methods enable powerful mechanisms for vulnerability detection at the cost of execution time overhead, because of the additional code that is inserted in the application program. Simulation has also been proposed for vulnerability testing, where a simulation environment is used to inject faults to a program and check its behavior [Du02]. This is a systematic method, but it is limited to input patterns that may cause errors.

Fuzz testing (fuzzing) provides an alternative, reliable approach with successful results and advantages over the previous methods. Fuzzing is a testing method that applies test inputs (vectors) to a system under test (SUT) and observes its outputs, as shown in Fig. 7.1. The goal of the fuzzer is to identify faults in the SUT, e.g., to detect inputs that lead to a system crash. The effectiveness of the fuzzer is based on its ability to identify as many vulnerabilities as possible covering effectively the input value space. If there is inability to identify whether a system or a program has crashed during a test, the effectiveness of the fuzzer cannot be evaluated.

Fuzzing provides several advantages over static and dynamic analysis. First, it can be applied to programs whose source code is not available. Second, it is independent of the internal complexity of the tested software which limits in practice other methods, such as static analysis. Because of this independence, the same fuzzing tool can be used to test similar programs independently of the programming language used for their coding. Finally, the identified faults and errors can be directly associated to the user input and can be evaluated easier.

Fuzz testing has its limitations. The space of input values is vast, and thus, it is impossible to test large systems for all their potential input values within reasonable time frames. A fuzzer that produces random input values can discover faults and vulnerabilities, but, in general, it will not detect easily many important vulnerabilities unless it follows some specific strategic approach. Its effectiveness depends on its ability to identify representative input values, which may originate from attacks or common errors with invalid inputs, and detect vulnerabilities that are useful to attackers.

Fuzz testing can be classified in three (3) categories, depending on the information that is available for the system under test (SUT) [Tak08] [Sut07], as shown in Fig. 7.1:

Fig. 7.1 Fuzz testing configuration

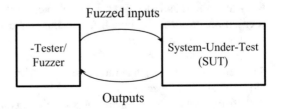

- White-box testing: the source code or the specification of the SUT is known.
- Black-box testing: the internal structure of the SUT is unknown –testing is limited to observations of SUT inputs and outputs.
- Grey-box testing: partial information for the SUT internal structure is available, e.g., through reverse engineering or static analysis results.

7.2.1   White-Box Fuzzing

Modern white-box fuzz testing tools exploit the information about the system's internal structure using symbolic execution techniques or taint analysis to identify vulnerabilities. Symbolic execution replaces symbolic values in the source code or the program flow, in order to evaluate code execution paths [Cad13]. These techniques have been explored widely in efforts such as DART [God05], SAGE [God12], EXE [Cad06], and KLEE [Cad08]. Tools like AEG [Avg11] and CRAX [Hua12] combine symbolic execution with concrete execution, employing concolic testing [Sen05] to identify vulnerabilities that lead to control flow hijacking. Such tools have been very successful in fuzz testing of Windows and Linux applications [God12, Cad06]. The techniques have the advantage that they can explore all possible modes of applications, since they use the source code, and identify dead code. However, they cannot identify logic errors in programs and are unable to explore all execution paths in large programs with complex structures. Tools that use taint analysis identify potential attack points in programs by tracing tainted values and then fuzz the input values to these attack points [Sch10]. BuzzFuzz [Gan09] and TaintScope [Wan10] are two representative tools that exploit taint analysis techniques.

7.2.2   Black-Box Fuzzing

Black-box fuzzing techniques do not have any structural information about the system under test. Since testing requires application of inputs to the system and observation of its outputs, one of the most popular targets of black-box fuzzing is the implementation of communication protocols because they provide the first point of entry to systems and they typically implement some standard; so, our description is focused on protocols, although the techniques can be applied to application and system software in general.

There are two main approaches to generate fuzz testing inputs to protocols: (i) data generation and (ii) data mutation [Nal12, Tak08, Sut07]. Data generation techniques create input packets to a protocol implementation either randomly or with a systematic method that takes into account the specifications of the specific protocol. The contents of these packets may be completely random, or they may take into account the structure of the packets, i.e., their fields, and insert either random or special values in the fields, depending on various parameters, such as the system interface or a specific targeted operation. In this case, the specification of the protocol needs to be integrated in the fuzzer. Clearly, the effectiveness of the fuzzing process depends on the successful integration of the protocol specification in the fuzzer, since any problem in that integration may lead to limited or no coverage of a wide range of tests.

Mutation fuzzing creates the test inputs based on legal protocol packets. It takes as input the legal packets and changes (mutates) some of their data, e.g., specific fields, in order to create the test packets that are input to the system. This approach is especially useful in cases where the protocol is complex, because the fuzzer does not construct packets from scratch but uses known legal packets and mutates them. Thus, the fuzzer does not need to include the protocol specification, and the author of the fuzzer does not need to delve into the details of the protocol, thus avoiding the risk of misinterpretations and creation of inappropriate packets.

These two main approaches are coupled with techniques that choose the values that are used in the generated or mutated packets. The most common techniques are:

1. Random: generates of random values without any consideration of packet structure, legal values, etc. The technique is fast, low cost, and quite successful [Mil90, Mil95, Mil06] but limited because it is characterized by low test coverage.
2. Block-based: manages data values in blocks, taking into account the specifications of protocols and creating meaningful blocks of values, in contrast to random values. The technique has been used widely in frameworks and tools, such as Spike [Ait02], SNOOZE [Ban06], Sulley [Ami14], Peach [Pea14], Autodafè [Vua06], and AspFuzz [Kit10], and is especially useful in mutation fuzzing. The success of the technique depends on the successful integration of protocol specs in the fuzzers.
3. Grammar-based: embeds a grammar in the fuzzer, in order to cover part of the specification of legal inputs to the system under test. Fuzzing inputs are created with the consideration of the grammar. PROTOS [PRO] is a representative tool using this technique.
4. Heuristic-based: generates new fuzzing inputs taking into account the effectiveness of the inputs applied in the past. Processing of the outputs obtained from the prior tests can be done with various methods such as with appropriate genetic algorithms [Spa07] or statistical analysis [Zha11].

There exist also approaches that construct protocol descriptions or specifications by observing real protocol traffic. With this information, related tools can make more effective decisions about how to mutate observed packets, in order to increase the effectiveness of mutation fuzzers. General Purpose Fuzzer (GPF) [Vda14] and AutoFuzz [Gor10] are representative tools that employ this approach. Interestingly, in mutation fuzzing there is also the approach of creating test cases based on existing attack traffic [Ant12, Tsa12].

7.3   Fuzzing Industrial Control Network Systems

Fuzz testing for industrial networks has attracted significant interest in the market and in academia, considering the increasing adoption of industrial control systems in critical infrastructures. Many commercial and open source fuzzing tools support industrial protocols. Sulley [Dev07] provides fuzzing modules for ICCP, Modbus, and DNP3 since 2007. ProFuzz [Koc], a fuzzing tool based on Scapy [Bio], supports fuzzing in PROFINET. Achilles test platform [Ach17] supports fuzzing for SCADA protocols, like Modbus/TCP and DNP3.

There is also research work in fuzzing industrial protocols using various techniques. Black-box mutation fuzzing, for example, has been explored for SCADA networks without any knowledge about the networking protocol [Sha11] and using the LZ-Fuzz tool [Bra08] to evaluate its effectiveness. OPC-MFuzzer [Wan13, Qi14] is a mutation fuzzer (based on Peach [Pea14]) for OPC SCADA fuzzing. Based on three different mechanisms to produce fuzzing inputs, the tool identified and confirmed known vulnerabilities that had been included previously in the National Vulnerability Database (NVD) [Nis].

Modbus fuzzing has attracted significant attention as well. BlackPeer [Byr06] produces inputs and checks outputs using a grammar that is included in the tool; although successful, it has limited flexibility as it cannot adjust easily to new tests. Sulley [Dev07], a block-based framework, enables methodical and easy mutation fuzzing through its Modbus module; however, its block-based approach is limited for testing devices that deviate from the standard implementation and are customized by the users. A framework for fuzz testing Modbus for security has also been proposed based on Scapy [Kob07].

7.4   Fuzzing Modbus

7.4.1   The Modbus Protocol

Modbus is an application protocol for industrial control system communication, which has become a standard published by Modbus IDA [Mod, ModS]. Its specification defines the protocol for direct communication over serial links as well as communication over TCP connections. The popular Modbus protocol stacks are shown in Fig. 7.2; it should be noted that, in correspondence with the ISO protocol reference model, Modbus is an application layer protocol defined to interface directly to layer 1 (serial) and layer 2 (HDLC) protocols –stacks (a) and (b) in the figure – or to TCP through an adjusting sublayer that is denoted as Modbus messaging (mapping) on TCP, as shown in stack (c).

The protocol implements client/server (alternatively, master/slave) communication through a request-response model between a control center and field devices, such as a SCADA and PLCs. For example, a SCADA master unit (client) may

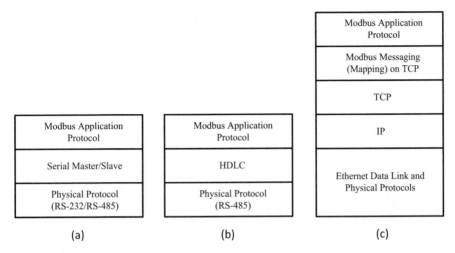

	Modbus Application Protocol
	Modbus Messaging (Mapping) on TCP
	TCP

Modbus Application Protocol	Modbus Application Protocol	IP
Serial Master/Slave	HDLC	Ethernet Data Link and Physical Protocols
Physical Protocol (RS-232/RS-485)	Physical Protocol (RS-485)	

(a)   (b)   (c)

Fig. 7.2 Modbus protocol stacks

Fig. 7.3 Modbus application packet

| Function code (FC) | Data |

request the reading of a sensor attached to a slave PLC (server), or it may request the writing of a command to an actuator to turn a switch.

Modbus application packets are simple, composed of two fields, a function code (FC) and data, as shown in Fig. 7.3. Requests from servers send the function code that defines the operation to be performed and the related data, e.g., an address or command. A response from a client includes the function code that was executed at the client and the resulting related data. Since an operation may not be successfully executed at the client, the protocol defines that the client will respond with the original function code if the related operation is executed correctly, or it will send an exception code indicating that the operation was not executed.

Modbus has three different classes of function codes: public codes, user-defined codes and reserved ones. Public codes are defined by the standard and include numbering and operation definition. Reserved codes are also public, but they cannot be used freely, since they have been defined and reserved for interoperability purposes with legacy industrial control systems. User-defined codes are available to developers and users to implement specialized function codes at will. Since the function code field is 8 bits, function codes can have 256 values, in the range 0–255. Public codes are in the ranges 1–64, 73–99, and 111–127; these ranges include the reserved codes. User-defined codes may have values in the ranges 65–72 and 100–110. The codes 128–255 are used to indicate errors; each function code has its unique related exception code, which differs from the function code at the most significant bit; with the 8-bit format, all function codes have "0" as their most significant bit and all exception codes have it as "1."

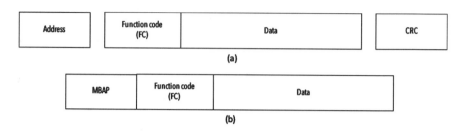

Fig. 7.4 Encapsulated Modbus application packets

Most Modbus function codes perform read and write operations to device data. For this purpose, Modbus considers that devices store data in tables. There are four different table types, based on the data entry size (1 bit or 16 bits) and the access operation allowed (read or read/write). The tables are denoted as (i) discrete input, with 1-bit entries and only read operations allowed; (ii) coils, with 1-bit entries and read/write operations allowed; (iii) input registers, with 16-bit entries and only read operations allowed; and (iv) holding registers, with 16-bit entries and read/write operations allowed. All four types of tables can have up to 64 K entries. Importantly, these tables are actually virtual, meaning that they can be physically separate in the device's memory or they can overlay over the same physical memory cells. Modbus can also access files, which are sequences of records (up to 10,000), and each record has a length measured with 16-bit units.

Modbus application packets (protocol data units, or PDUs) are encapsulated in lower layer protocol packets to be transmitted. When serial connections are used, the application packets are encapsulated by the data link control (DLC) protocol and produce DLC PDUs that are then transmitted by the serial protocol. DLC packets add an address field for the slave next to the function code field and a checksum next to the data field of the application protocol, as Fig. 7.4a shows. In the case of the serial physical layer, there are two formats for the DLC packets, denoted as RDU and ASCII. The main difference between the two is the size of the slave address and the size of the function code field: in RTU format, they are both one byte, while in the ASCII format, each one is 2 bytes long.

Modbus over TCP is performed by extending the Modbus application packet first with an additional header, named MBAP (Modbus Application Protocol) header as shown in Fig. 7.4b, and then encapsulating this extended packet by the TCP/IP protocol stack, which employs Ethernet at the data link control and physical protocol layers, as shown in Fig. 7.2c.

Modbus does not include security mechanisms such as authentication, confidentiality, or integrity. The lack of security renders its implementations vulnerable to a wide range of attacks. The lack of confidentiality enables attackers to extract information from captured packets, while the lack of integrity checks does not allow a receiver of a packet to identify whether the packet has been altered. Replay attacks are possible as well and the lack of non-repudiation mechanisms can lead to inability to analyze and audit systems credibly.

7.4.2   Modbus/TCP Fuzzer

There exist several Modbus fuzzers, as described in Sect. 7.2. In this subsection, we present the approach and results of MTF (Modbus/TCP fuzzer) [Voy15] as a representative example. The choice of MTF is based on its characteristics that show the trends in fuzzing technology today: it is an automated tool, it provides good coverage of input tests, and it does not require physical access to the system under test, operating remotely over the network. These characteristics make MTF an attractive tool for testing security and compliance of Modbus connected devices.

MTF incorporates the specification of Modbus/TCP and supports fuzzing both master and slave devices on the network. As an automated tool for fuzzing, MTF operates in three main phases: (i) reconnaissance, (ii) attack, and (iii) failure detection. In the first phase, MTF identifies the operational characteristics and parameters of the tested system. In the second phase, it applies tests to the system and collects its responses, while in the third phase it evaluates the collected (observed) responses to identify security problems and system failures.

Reconnaissance is an important operation in automated black-box or gray-box fuzzers, because it identifies the operations performed by the system under test and its important parameters. In the case of Modbus, in order to generate meaningful tests, one needs to know the function codes used by the system as well as its memory model, i.e., the four memory types – discrete inputs, coils, input registers, and holding registers – that are specified by the standard. MTF explores the function codes through different methods, in order to accommodate different types of devices that may be fully or partially conformant with the standard. A straightforward method is to ask the device for identification information – the standard specifies function code 43 for this operation – and then, based on this, to find information off-line about the supported function codes, e.g., from a manual. Alternatively, it sends legitimate requests and examines the responses, which indicate whether the requests have been executed or not (as described in the standard specification), or it monitors traffic from the device and extracts functional information from that.

In regard to the memory model of the tested system, MTF effectively identifies the boundary memory addresses for each type of memory. This is done either actively, sending packets with the appropriate function codes probing specific address values, or passively, observing traffic which eventually indicates memory bounds, although these bounds may be approximate.

Taking into account the list of function codes and the memory mapping for the four memory types, the fuzzer can construct legitimate packets and fuzz them in order to test the system. Since the supported function codes are known, MTF constructs a set of packet sequences for each supported function code, where each sequence implements a potential attack to the system; such attacks include packet removal, packet injection, and packet field manipulation.

Packet field manipulation is performed with field values that are boundary, random, or illegal.

When tests are applied, the response, or its absence, is recorded. The tool records the sequence of all tests and related responses and produces a list of errors which are invalid responses (out of specification), valid but with incorrect parameters (values, size, etc.), and delayed or incomplete (no response). Further processing of the records, including both the valid request/response pairs and the errors, leads to detection of security and dependability problems, i.e., malicious or accidental failures.

The MTF approach is representative of the trends in fuzzing industrial protocols. It provides a complete approach to fuzzing, starting with reconnaissance, continuing with meaningful tests and, finally, analyzing the results for security and reliability failures. Its practicality has been demonstrated through the prototype implementation described in the original work [Voy15], which has been used to evaluate several commercial and open source Modbus subsystems and for several attacks. The attacks include packet dropping, packet injection, illegal field values, altered function codes, and even flooding, leading to denial of service attacks. Importantly, many of these attacks have been successful against commercial Modbus implementations, as the reported original results demonstrate. Interestingly, MTF succeeds in attacking these implementations much more efficiently than alternative tools, i.e., with a significantly smaller number of packets. Overall, the results demonstrate that the approach of generation fuzzing is an effective and efficient fuzzing method.

References

[Ach17]   Wurldtech- GE Digital, Achilles Test Platform, 2017. https://www.ge.com/digital/sites/default/files/achilles_test_platform.pdf

[Ait02]   Aitel, D. (2002). An introduction to SPIKE, the Fuzzer Creation Kit. Presented at The BlackHat USA Conference. www.blackhat.com/presentations/bh-usa-02/bh-us-02-aitel-spike.ppt

[Ami14]   Amini, P. (2014). Sulley: Pure Python fully automated and unattended fuzzing framework. https://github.com/OpenRCE/sulley

[Ant12]   Antunes, J., & Neves, N. (2012). Recycling test cases to detect security vulnerabilities. In *Proceedings of the 23rd International Symposium on Software Reliability Engineering*, Dallas, Texas, November 27–30, 2012, pp. 231–240.

[Avg11]   Avgerinos, T., Cha, S. K., Hao, B. L. T., & Brumley, D. (2011). AEG: Automatic Exploit Generation. In *Proceedings of the Network and Distributed System Security Symposium (NDSS'11)*, San Diego, California, February 6–9, 2011.

[Ban06]   Banks, G., et al. (2006). SNOOZE: toward a Stateful NetwOrk prOtocol fuzZEr. In *Proceedings of the 9th Information Security Conference (ISC '06)*, pp. 343–358.

[Bio]   Biondi, P. Scapy, python interactive packet manipulation framework. http:/www.secdev.org/projects/scapy/

[Bra08]   Bratus, S., Hansen, A., & Shubina, A.(2008). LZFuzz: A fast compression-based Fuzzer for poorly documented protocols. Technical Report TR2008–634, Dept. of Computer Science, Dartmouth College, New Hampshire.

[Byr06]   Byres, E. J., Hoffman, D., & Kube, N. (2006). On shaky ground – A study of security vulnerabilities in control protocols. In *Proceedings of the 5th International Topical Meeting on Nuclear Plant Instrumentation Controls, and Human Machine Interface Technology*, American Nuclear Society, Albuquerque, November 12–16, 2006.

[Cad06] Cadar, C., Ganesh, V., Pawlowski, P., Dill, D., & Engler, D. (2008). EXE: Automatically generating inputs of Death. In: Proceedings of CCS'06, Oct–Nov 2006 (extended version appeared in ACM TIS-SEC 12:2, 2008).

[Cad08] Cadar, C., Dunbar, D., & Engler, D. (2008). KLEE: Unassisted and automatic generation of high-coverage tests for complex systems programs. In *Proceedings of OSDI'08*, December 2008.

[Cad13] Cadar, C., & Sen, K. (2013). Symbolic execution for software testing: Three decades later. *Communications of the ACM, 56*(2), 82–90.

[Cla07] Clause, J., Li, W., & Orso, A. (2007). Dytan: A generic dynamic taint analysis framework. In *Proceedings of the 2007 International Symposium on Software Testing and Analysis (ISSTA'07)*, London, UK, July 9–12, 2007, pp. 196–206.

[Che07] Chess, B., & West, J. (2007). *Secure programming with static analysis*. USA: Pearson Education.

[Cow98] Cowan, C., et al. (1998). StackGuard: Automatic adaptive detection and prevention of buffer-overflow attacks. In *Proceedings of the 7th Usenix Security Symposium*, San Antonio, Texas, January 26–29, 1998.

[Dev07] Devarajan, G. (2007). Unraveling SCADA protocols: Using Sulley Fuzzer. Presented at the DefCon'15 Hacking Conference, 2007.

[Du02] Du, W., & Mathur, A. P. (2002). Testing for software vulnerability using environment perturbation. *Quality and Reliability Engineering International, 18*(3), 261–272.

[Gan09] Ganesh, V., Leek, T., & Rinard, M. (2009). Taint-based directed whitebox fuzzing. In *Proceedings of the 31st International Conference on Software Engineering (ICSE'09)*, Vancouver, Canada, May 16–24, 2009, pp. 474–484.

[God05] Godefroid, P., Klarlund, N., & Sen, K. (2005). DART: Directed Automated Random Testing. In *Proceedings of the 2005 ACM SIGPLAN Conference on Programming language design and implementation*, Chicago, USA, June 12–15, 2005, pp. 213–223.

[God12] Godefroid, P., Levin, M. Y., & Molnar, D. (2012). SAGE: Whitebox fuzzing for security testing. *ACM Queue, 10*(1).

[Gor10] Gorbunov, S., & Rosenbloom, A. (2010). Autofuzz: Automated network protocol fuzzing framework. *IJCSNS, 10*(8), 239–245.

[Hua12] Huang, S. K., Huang, M. H., Huang, P. Y., Lai, C. W., Lu, H. L., Leong, W. M. (2012). CRAX: Software crash analysis for automatic exploit generation by modeling attacks as symbolic continuations. *IEEE 6th International Conference on Software Security and Reliability*, June 20–22, 2012, pp. 78–87.

[Kit10] Kitagawa, T., Hanaoka, M., & Kono, K. (2010). AspFuzz: A state-aware protocol fuzzer based on application-layer protocols. In *Proceedings of the IEEE Cymposium on Computers and Communications*, Italy, 2010, pp. 202–208.

[Kob07] Kobayashi, T. H., Batista, A. B., Brito, A. M., & Motta Pires, P. S. (2007). Using a packet manipulation tool for security analysis of industrial network protocols. In *Proceedings of 2007 IEEE Conference on Emerging Technologies and Factory Automation*, Patras, 2007, pp. 744–747.

[Koc] Koch, R. Profuzz. https://github.com/HSASec/ProFuzz

[Mil90] Miller, B. P., Fredriksen, L., & So, B. (1990). An empirical study of the reliability of UNIX utilities. *Communications of the ACM, 33*(12), 32–44.

[Mil95] Miller, B. P., et al. (1995). Fuzz revisited: A re-examination of the reliability of UNIX utilities and services. Technical report TR-1268, Department of Computer Sciences, University of Wisconsin-Madison.

[Mil06] Miller, B. P., Cooksey, G., & Moore, F. (2006). An empirical study of the robustness of MacOS applications using random testing. In *Proceedings of the 1st International Workshop on Random testing*. Portland, Maine, July 20, 2006, pp. 46–54.

[Mod]   ModBus Organization. ModBus Application Protocol Specification http://www.modbus.org/docs/ModbusApplication/ProtocolV11b.pdf

[ModS]   Modbus Serial Line Protocol and Implementation Guide V1.02 (Modbus_over_serial_line_V1_02.pdf).

[Nal12]   McNally, R., Yiu, K., Grove, D., & Gerhardy, D. Fuzzing: The State of the Art. Technical Note DSTO-TN-1043, Defence Science and Technology Organization, Australia, 02–2012.

[New04]   Newsome, J., & Song, D. Dynamic taint analysis for automatic detection, analysis, and signature generation of exploits on commodity software. Technical report CMU-CS-04-140, 2004 (revised 2005).

[Nis]   http://nvd.nist.gov

[Pea14]   Peach Fuzzing Platform, http://www.peach.tech/products/peach-fuzzer/, 2017.

[PRO]   PROTOS-Security Testing of Protocol Implementations. http//www.ee.oulu.fi/roles/ouspg/Protos/

[Qi14]   Qi, X., Yong, P., Dai, Z., Yi, S., & Wang, T. (2014). OPC-MFuzzer: A novel multi-layers vulnerability detection tool for OPC protocol based on fuzzing technology. *International Journal of Computer and Communication Engineering, 3*(4), 300–305.

[Sch10]   Schwartz, E. J., Avgerinos, T., & Brumley, D. (2010). All you ever wanted to know about dynamic taint analysis and forward symbolic execution (but might have been afraid to ask). *2010 IEEE Symposium on Security and Privacy*.

[Sen05]   Sen, K., Marinov, D., & Agha, G. (2005). CUTE: A concolic unit testing engine for C. In *Proceedings of the 10th European Software Engineering Conference (held jointly with 13th ACM SIGSOFT International Symposium on the Foundations of Software Engineering)*, September 5–9, 2005, pp. 263–272.

[Sfo]   http://www.securityfocus.com

[Sha11]   Shapiro, R., Bratus, S., Rogers, E., & Smith, S. (2011). Identifying vulnerabilities in SCADA systems via fuzz-testing. *Critical Infrastructure Protection V, IFIP AICT, 367*, 57–72.

[Spa07]   Sparks, S., Embleton, S., Cunningham, R., & Zou, C. (2007). Automated vulnerability analysis: Leveraging control flow for evolutionary input crafting. In *Proceedings of the 23rd Annual IEEE Computer Security Applications Conference (ACSAC 2007)*, pp. 477–486.

[Str]   http://www.securitytracker.com

[Sut07]   Sutton, M., Greene, A., & Amini, P. (2007). *Fuzzing: Brute force vulnerability discovery*. Addison-Wesley Professional.

[Tak08]   Takanen, A., DeMott, J., & Miller, C. (2008). *Fuzzing for software security testing and quality assurance*.

[Tsa12]   Tsankov, P., Torabi Dashti, M., Basin, D. (2012). SECFUZZ: Fuzz-testing security protocols. In *Proceedings of the 7th International Workshop on Automation of Software Test (AST 2012)*, June 2–3, 2012, Zurich, Switzerland.

[Vda14]   VDA Labs, "General Purpose Fuzzer." Rockford, Michigan, 2014, www.vdalabs.com/tools/efs gpf.html

[Vie00]   Viega, J., et al. (2000). ITS4: A static vulnerability scanner for C and C++ code. In *Proceedings of 16th Annual IEEE Conference Computer Security Applications (ACSAC'00)*, New Orleans, Louisiana, 2000, pp. 257–267.

[Voy15]   Voyiatzis, A. G., Katsigiannis, K., & Koubias, S. (2015). A Modbus/TCP Fuzzer for testing internetworked industrial systems. In *Proceedings of the 20th IEEE International Conference on Emerging Technologies and Factory Automation (ETFA 2015)*. Luxembourg, September 8–11, 2015, pp. 1–6.

[Vua06]   Vuagnoux, M. (2006). Autodafe: An Act of Software Torture. Swiss Federal Institute of Technology (EPFL), Cryptography and Security Laboratory (LASEC). http://autodafe.sourceforge.net

[Wan10]   Wang, T., Wei, T., Gu, G., & Zou, W. (2010). TaintScope: A checksum-aware directed fuzzing tool for automatic software vulnerability detection. *2010 IEEE Symposium on Security and Privacy*, Oakland, CA, USA, 2010, pp. 497–512.

[Wan13]   Wang, T., et al. (2013). Design and implementation of fuzzing technology for OPC pro-
tocol. In *Proceedings of 9th International Conference on Intelligent Information Hiding and
Multimedia Signal Processing*, Beijing, China, 2013, pp. 424–428.
[Zha11]   Zhao, J., Wen, Y., & Zhao, G. (2011). H-fuzzing: A new heuristic method for fuzz-
ing data generation. In *Proceedings of Network and Parallel Computing*, LNCS, Vol. 6985,
Springer, 2011, pp. 32–43.

Index

© Springer International Publishing AG 2018
D. Serpanos, M. Wolf, *Internet-of-Things (IoT) Systems*,
https://doi.org/10.1007/978-3-319-69715-4

Printed in the United States
By Bookmasters